C0-ALU-724

Ring Nitrogen and Key Biomolecules

Ring Nitrogen and Key Biomolecules

The biochemistry of *N*-heterocycles

E.G. Brown

Professor of Biochemistry, School of Biological Sciences,
University of Wales Swansea, UK

KLUWER ACADEMIC PUBLISHERS
DORDRECHT / BOSTON / LONDON

Library of Congress Cataloging in Publication Card Number: 98-70669

ISBN 0 412 83570 3 (HB) 0 412 62980 1 (PB)

Published by Kluwer Academic Publishers,
P.O. Box 17, 3300 AA Dordrecht, The Netherlands.

Sold and distributed in North, Central and South America
by Kluwer Academic Publishers,
101 Philip Drive, Norwell, MA 02061, U.S.A..

In all other countries, sold and distributed
by Kluwer Academic Publishers Group,
P.O. Box 322, 3300 AH Dordrecht, The Netherlands.

Printed in Great Britain

Contents

Abbreviations

Abbreviations used are those recommended by the IUPC-IUB Joint Commission on Biochemical Nomenclature. In some of the more complex diagrams and formulae, 'P' has been used to represent an orthophosphate group.

Introduction

> 'The important thing in science is not so much to obtain new facts as to discover new ways of thinking about them'
>
> *Sir W L Bragg (1968)*

A cursory glance in a contemporary biochemistry textbook immediately reveals the omnipresence and essentiality of *N*-heterocyclic molecules in living organisms. These nitrogen-containing ring structures range from the relative structural simplicity of ATP to the heteropolymeric complexity of the nucleic acids; they include the majority of coenzymes, metabolic regulators and integrators such as adenosine and GTP, signalling compounds like the cyclic nucleotides and the plant cytokinins, and biochemically functional pigments, of which haemoglobin, the cytochromes, and chlorophyll are examples. In addition to compounds of known biological importance there are numerous naturally occurring *N*-heterocyclic molecules of unknown significance. Little is known, for example, about possible biological functions of alkaloids in the plants synthesizing them, yet many of these compounds have high commercial value as pharmaceuticals.

The prime objective of this book is to collate and to integrate current knowledge of biologically important *N*-heterocyclic compounds, not only to provide a compendium for advanced students and researchers in biochemistry, and to those in the interfacing subject areas of chemistry, biology and pharmacology, but also to stimulate interest in the relationship between chemical structure and physiological function within this key group of compounds.

Consideration of the role of biochemically important compounds by juxtaposing them in families of related chemical structure, rather than of related biological function, is not a stance commonly adopted by authors in biochemistry but it has much to commend it. It highlights fascinating and fundamental biochemical and chemical questions, and points to new ways of looking at physiological function. What is it, for example, that makes the uridine nucleotide UDP so particularly suited to transporting aldoses, such as glucose and galactose, whereas the cytidine nucleotide CDP primarily transports alcohols like ribitol, glycerol, and ethanolamine? Why are ribonucleotides so intimately connected with the regulation and integration of metabolism, and with intracellular signalling? Are these nucleotide-regulated functions related in some way to the informational content of nucleic acids? Where do the plant cytokinins, which are also purines, fit into this picture? What is the biochemistry underlying the presence of either a purine or pyrimidine component within the molecular structure of the majority of coenzymes? These and a host of other important questions present themselves readily; the answers are a little more elusive.

The impact of molecular biology on biochemical research has tended to displace the teaching of biochemistry towards biology and resulted in a diminution of interest in biological chemistry. Meanwhile, most university chemistry courses maintain their traditional reluctance to consider biological systems. Consequently, as biochemistry is essentially the synergistic interaction of biology and chemistry, its hitherto highly successful approach to understanding life processes is in danger of being eroded. A second objective of the book has therefore been to provide, for both the biologist and the chemist, a biochemical survey in a chemical framework which illustrates the importance of the **biochemical** approach to biological problems.

Why single out *N*-heterocyclic compounds for this survey? Their numerous key catalytic roles in metabolism means that they dominate the biochemical scene in all cells and tissues; there are few biochemical reaction sequences that do not involve a *N*-heterocyclic compound as substrate, product, or coenzyme. In fact, it could be said that in the first place nature selected these compounds, and they are the hub of metabolism. Molecules such as the pyridine nucleotides NAD and NADP, the adenine nucleotides ATP and cyclic AMP, coenzyme A, and the master genetic molecules of DNA and RNA, are all *N*-heterocyclic compounds linking, coordinating and controlling metabolism. No other group of compounds cuts such a swathe through biochemical organization.

What is there about the structure and properties of *N*-heterocycles that has led to their selection by nature for these critical roles? Merely to consider this question is to see a fresh perspective on the chemistry of life. This is to be recommended not only to the academic scientist but to the pharmacologist, agricultural chemist and food scientist who, in turn, exploit biochemical knowledge in their respective fields of interest and application.

GENERAL READING

Gilchrist, T.L. (1997) Heterocyclic chemistry (3rd Edn.).

Hurst, D.T. (1980) An introduction to the chemistry and biochemistry of pyrimidines, purines, and pteridines.

Mathews, C.K. and Van Holde, K.E. (1995) Biochemistry (2nd Edn.)
Pozharskii, A.F., Soldatenkov, A.T. and Katritzky, A.R. (1997) Heterocycles in Life and Society.
Sinnott, M. (ed) (1977) Comprehensive Biological Catalysis vol. 1–4, Academic Press, London.
Stryer, L. (1995) Biochemistry (4th Edn.)
Zubay, G.L., Parson, W.W. and Vance, D.E. (1995) Principles of Biochemistry.

CHAPTER 1

Pyrroles and tetrapyrroles, including porphyrins

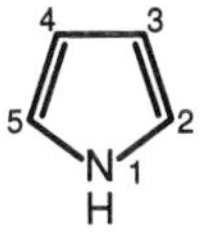

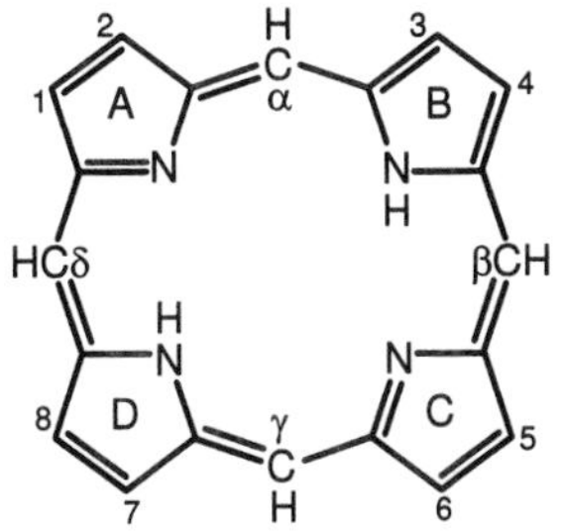

1.1 INTRODUCTION

Whereas the monopyrroles are of little direct biochemical interest except as metabolic intermediates, the tetrapyrroles constitute a major group of fundamental importance of which the porphyrins, such as haem and chlorophyll, are examples. To appreciate fully the biochemistry of the tetrapyrroles, some knowledge of the salient features of pyrrole chemistry helps and an outline is given below in Section 1.3.

1.2 NATURAL OCCURRENCE OF MONOPYRROLES

Pyrrole itself is a colourless liquid and has been known since the middle of the last century. Only a few simple pyrrole derivatives occur naturally. They include an antifungal antibiotic, pyrrolnitrin (1), produced by certain strains of *Pseudomonas*, another antibiotic called pyoluteorin (2), and a metabolite known as porphobilinogen (3) which accumulates in the urine of patients with acute porphyria and has been shown to be the precursor of the porphyrins.

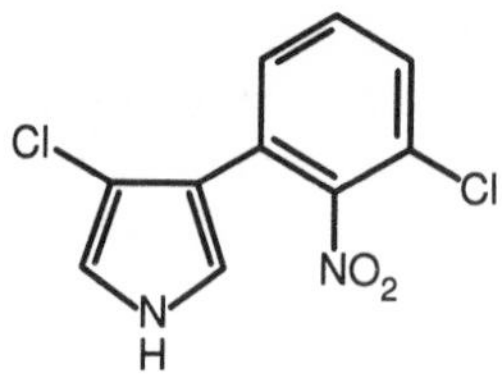

(1)

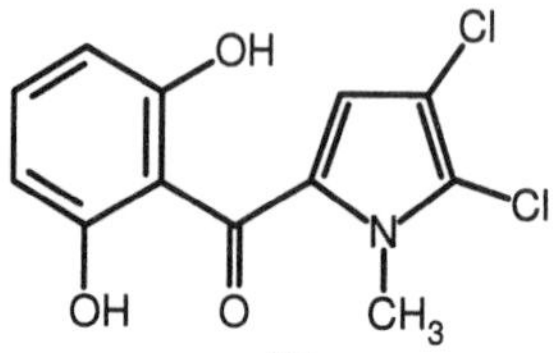

(2)

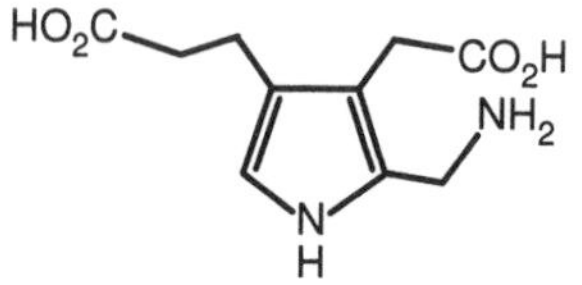

(3)

1.3 PYRROLIC PROPERTIES OF BIOCHEMICAL RELEVANCE

The chemical properties of pyrrole are characteristic of an aromatic system although over the years the extent of this aromaticity has been much debated. By comparison with typical secondary amines, pyrrole has negligible basicity and has been classified as a 'π-excessive' heteroaromatic compound. This π-excessive character, attributable to six electrons being distributed over 5 atoms, means that it is extremely susceptible to electrophilic attack. Generally the 2-position is thermodynamically favoured by such attack but under some conditions, the 3-position may be preferred. Presence of electron-donating substituents at the 2- and 3-positions promote electrophilic substitution at the 5- and 2-positions, respectively. If there is an electron-withdrawing substituent at the 2-position, electrophilic attack results in substitution at both the 4- and 5-positions.

Having no substituent at the *N*-atom, pyrroles are weakly acidic and treatment with strong acid yields anions capable of reaction at the *N*-atom with electrophiles. Their π-electron excessive character means that pyrroles are relatively inert with respect to nucleophilic addition or substitution reactions.

1.4 DISCOVERY AND NATURAL OCCURRENCE OF TETRAPYRROLES AND RELATED COMPOUNDS

1.4.1 Porphyrins

The first preparation of a porphyrin in a reasonable state of purity was made by the physician-chemist Thudichum in 1867. He described the action of strong acid upon blood and the separation from the mixture of a pigment that he called 'cruentine'. Some years later, the German physiological chemist Hoppe-Seyler renamed this substance haematoporphyrin (4). A related iron (FeIII)-containing chloride, haemin (5),

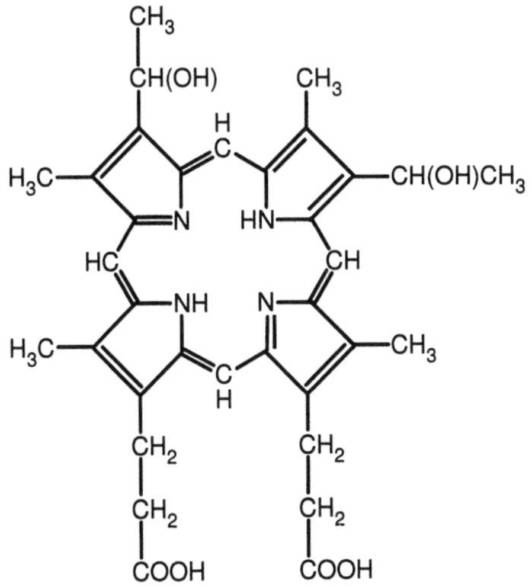

(4)

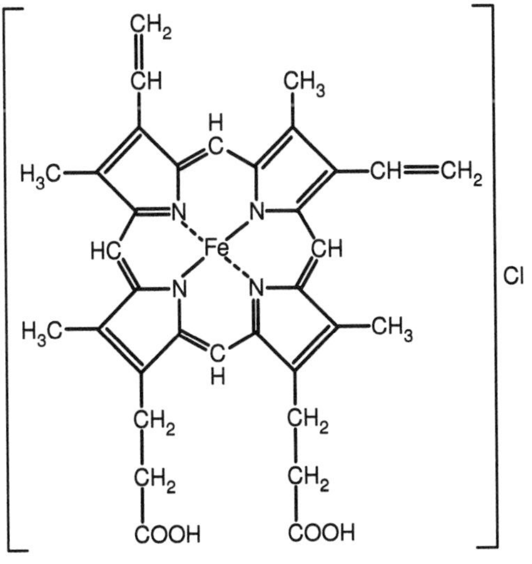

(5)

obtained by treating blood with hot acetic acid was prepared and investigated by Hans Fischer and it was from this study that an understanding of the chemistry of porphyrins developed.

The porphyrin molecule consists essentially of four pyrrole rings linked by methene bridges to form a macrocycle (6). Naturally occurring

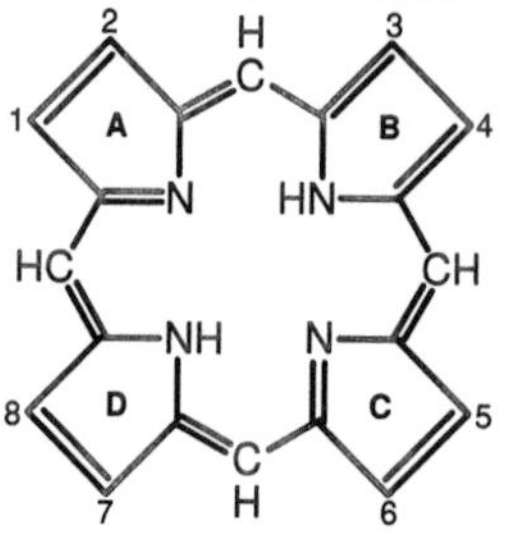

(6)

porphyrins differ from one another by the nature of the substituents at positions 1 to 8. The simplest structure, although not known to occur in nature, is one in which all the positions 1 to 8 are occupied by hydrogen atoms (6). This compound, synthesized by Fischer was given the name *porphin* and it was used by him as the basis for a rational system of porphyrin nomenclature. Removal of the iron (FeIII) from haemin, by strong acid under reducing conditions, yields protoporphyrin (7). This

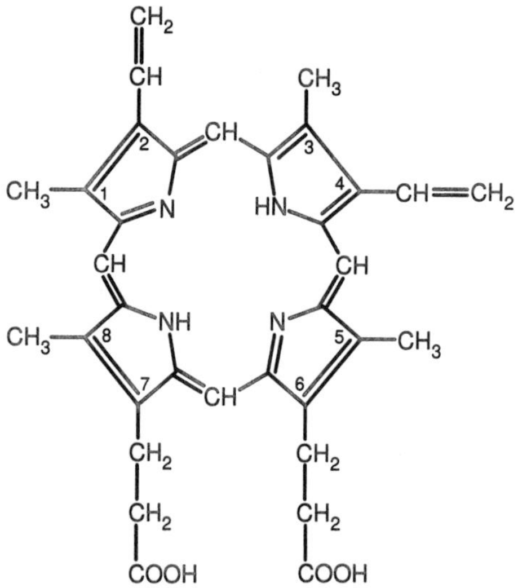

(7)

is where the structural problem begins for, excluding biologically atypical substitution patters, there are fifteen possible arrangements of the four methyls, two vinyls, and two propionic acid residues around the 8 positions on the porphin molecule. The problem confronting Fischer was, in essence, which group goes where? He solved the problem by the prodigious feat of synthesizing all fifteen isomers and comparing their properties with the natural product. It was his ninth isomeric sample that coincided and gave rise to the term protoporphyrin IX for the natural isomer.

To facilitate further the classification of porphyrins, Fischer synthesized all four isomers of a porphyrin in which the substituents at positions 1 to 8 consist of four methyl groups and four ethyl groups but in which no single ring possesses two identical substituents. These compounds, whilst not naturally occurring, were modelled on the substitution patterns of natural prophyrins. He designated his compounds **aetioporphyrins I to IV** and suggested that each of them can be regarded as the theoretical starting point for a series of porphyrins derived by inserting vinyl, acetic acid, propionic acid, etc., in place of the alkyl groups. The structures of the four aetioporphyrins are shown in Fig. 1.1. Comparison of the structure of protoporphyrin

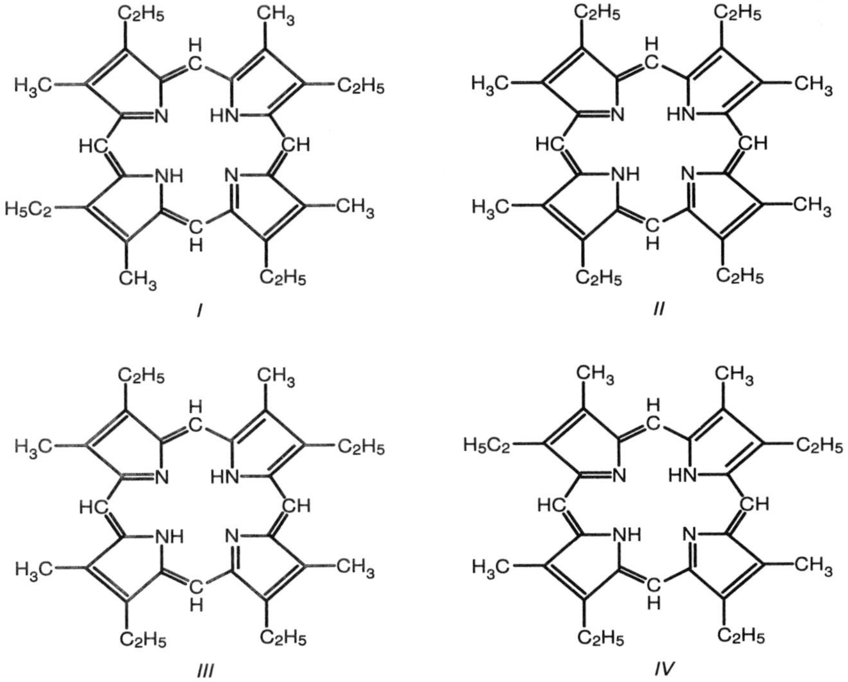

Fig. 1.1 The four isomeric aetioporphyrins synthesized by Fischer, and in which four methyl groups and four ethyl groups are present as substituents on the pyrrole rings, no one ring having two identical substituents. Each aetioporphyrin can be regarded as the theoretical starting point for a series of porphyrins derived by inserting vinyl ($-CH{=}CH_2$), acetic acid ($-CH_2COOH$), propanoic acid ($-CH_2CH_2COOH$) etc., in place of the alkyl groups. Comparison of the substitution pattern of a porphyrin with those above enables the compound to be allocated to one specific series (i.e. I, II, III or IV).

IX (7) with them indicates that it derives from aetioporphyrin III and it is therefore said to be a series III porphyrin. In fact, almost all known natural porphyrins belong to series I or III with the latter predominating by far.

The porphyrins are widely distributed in nature. Normal urine contains 10–200 μg l^{-1} of a mixture of coproporphyrins I and III and smaller amounts of uroporphyrins. Both coproporphyrin and protoporphyrin are present in erythrocytes where the main pigment is, of course, haemoglobin. This molecule is composed of a protein, globin, combined with haem, the iron (FeII) complex of protoporphyrin. The structures of the naturally occurring uroporphyrins and coproporphyrins are shown in Fig. 1.2.

Another biologically important group of haem proteins are the cytochromes, essential components of the respiratory and photosynthetic electron transport chains. Catalase and peroxidase, both enzymes using hydrogen peroxide as a substrate, are also haem proteins. The chlorophylls (Fig. 1.3) are magnesium complexes of modified porphyrins called pheophytins. Chlorophyll *a* is the predominant chlorophyll in nature and is present in all oxygen-producing photosynthetic organisms from algae to higher plants. Higher plants also contain chlorophyll *b*, as do some algae and mosses. Chlorophylls, c_1, c_2 and *d* are found in some species of algae. Most photosynthetic bacteria contain bacteriochlorophyll *a* as their major photosynthetic pigment whereas bacteriochlorophyll *b* is mainly found in some species of *Rhodopseudomonas*. In each of the chlorophylls and bacteriochlorophylls, the propionic acid residue attached to ring D is esterified with the diterpenoid alcohol phytol (8). Bacteriochlorophyll *a* differs from chlorophyll *a* in two respects. The 2-vinyl substituent of chlorophyll *a* is replaced in bacteriochlorophyll *a* by an acetyl group, and ring B of bacteriochlorophyll *a* is partially reduced (Fig. 1.3).

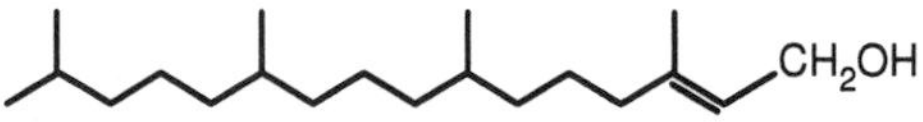

CH_2OH

(8)

Another interesting metalloporphyrin is turacin, the bright red pigment of the flight feathers of the African turaco bird. This pigment has been identified as copper uroporphyrin III and is noteworthy in two respects. First, it is the only known example of a naturally occurring copper porphyrin and second, it is the only reported occurrence of uroporphyrin III as an end product of metabolism.

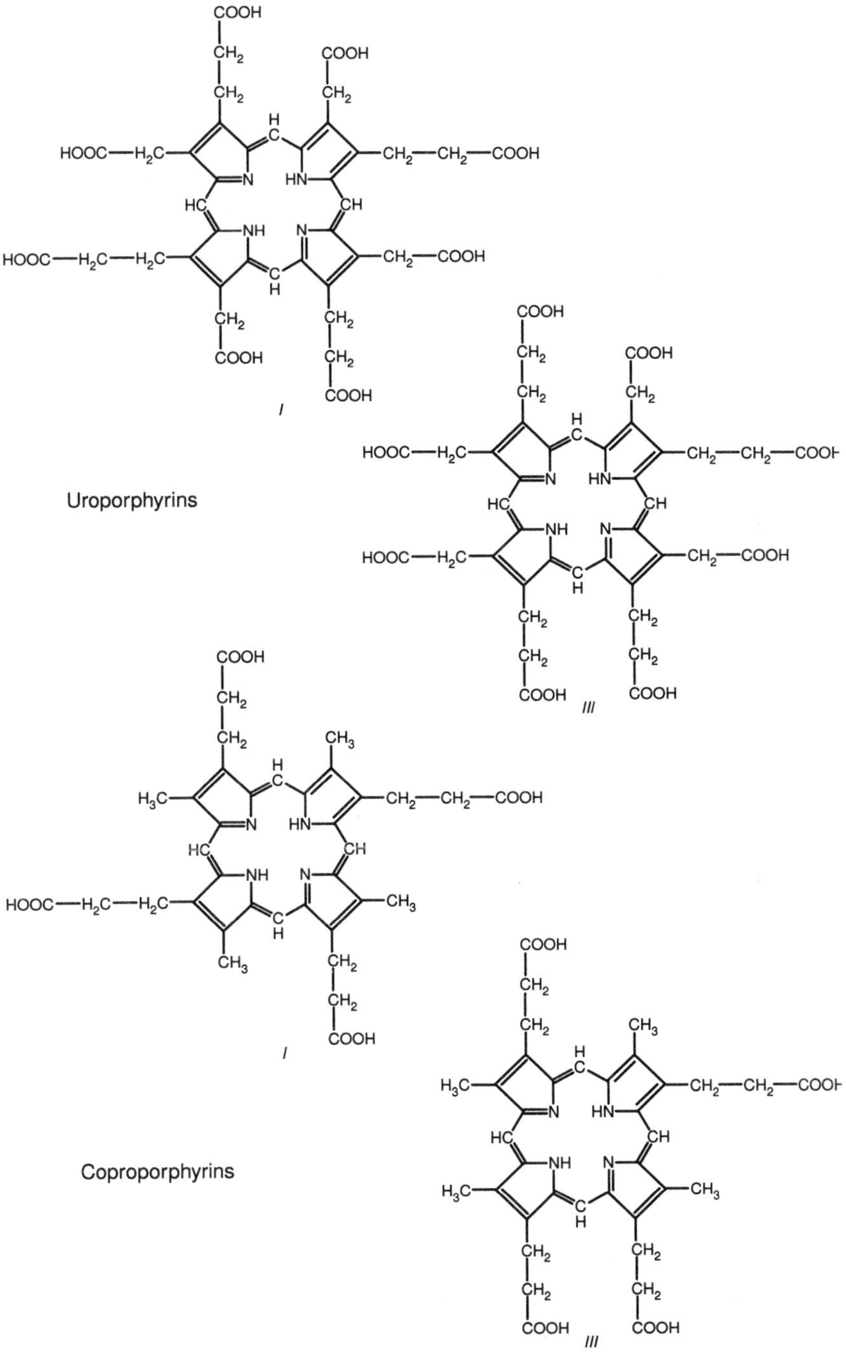

COOH
CH$_2$
CH$_2$
COOH
CH$_2$
H
C
HOOC—H$_2$C—
—CH$_2$—CH$_2$—COOH
N
HN
HC
CH
NH
N
HOOC—H$_2$C—H$_2$C—
—CH$_2$—COOH
C
H
CH$_2$
COOH
CH$_2$
CH$_2$
COOH
I

Uroporphyrins

COOH
CH$_2$
CH$_2$
COOH
CH$_2$
H
C
HOOC—H$_2$C—
—CH$_2$—CH$_2$—COOH
N
HN
HC
CH
NH
N
HOOC—H$_2$C—
—CH$_2$—COOH
C
H
CH$_2$
CH$_2$
COOH
CH$_2$
CH$_2$
COOH
III

COOH
CH$_2$
CH$_2$
CH$_3$
H
C
H$_3$C—
—CH$_2$—CH$_2$—COOH
N
HN
HC
CH
NH
N
HOOC—H$_2$C—H$_2$C—
—CH$_3$
C
H
CH$_3$
CH$_2$
CH$_2$
COOH
I

Coproporphyrins

COOH
CH$_2$
CH$_2$
CH$_3$
H
C
H$_3$C—
—CH$_2$—CH$_2$—COOH
N
HN
HC
CH
NH
N
H$_3$C—
—CH$_3$
C
H
CH$_2$
CH$_2$
COOH
CH$_2$
CH$_2$
COOH
III

Fig. 1.2 The naturally occurring uroporphyrins and coproporphyrins. Except in some pathological samples, series III isomers predominate.

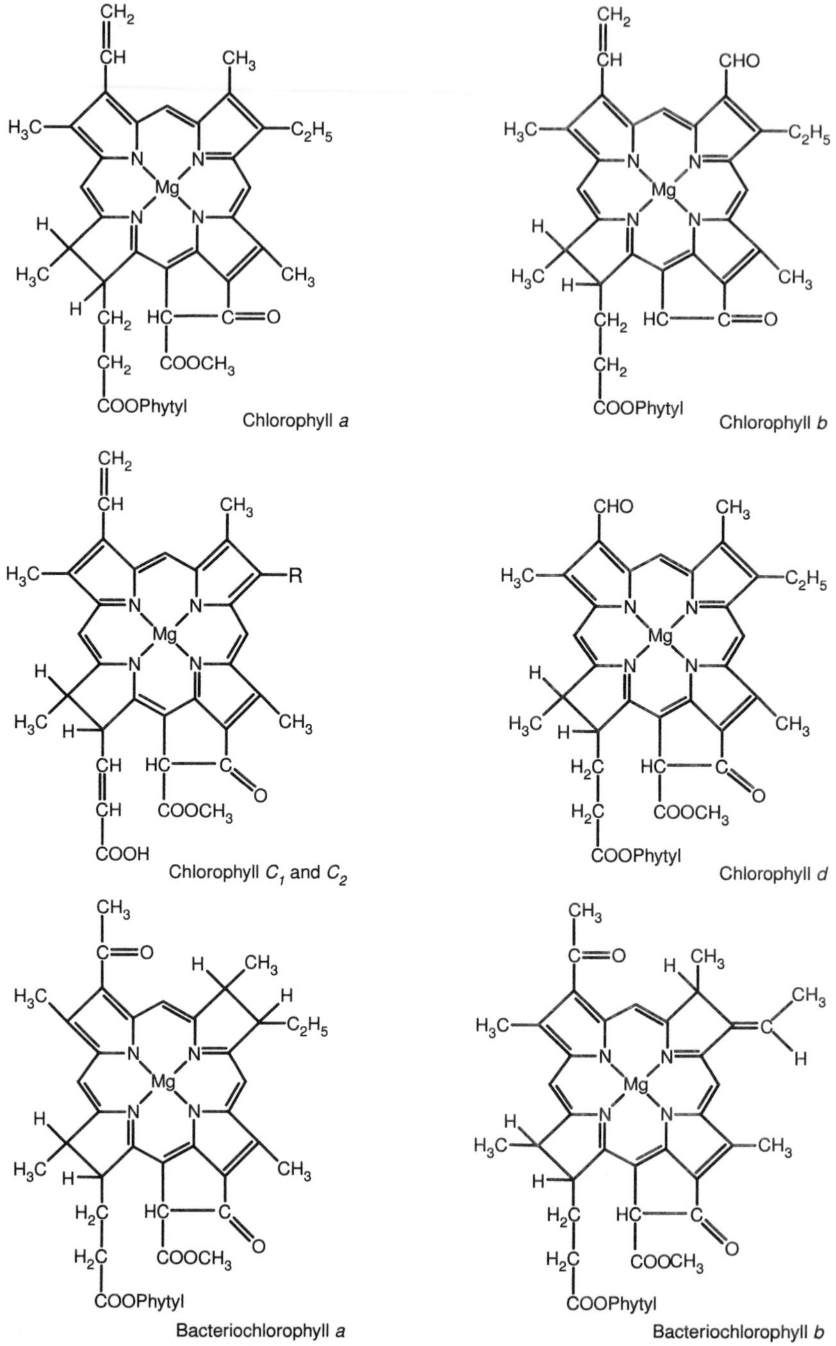

Fig. 1.3 Structures of the chlorophylls: chlorophyll c_1, R = $-CH{=}CH_2$; chlorophyll c_2, R = $-CH_2CH_3$.

1.4.2 Bile pigments and phycobilins

Bile pigments are linear tetrapyrroles arising from the catabolism of haem. They are exemplified by biliverdin (9) which is the green component of bile, and bilirubin (10) which is the main red constituent. The biological production of these compounds is outlined in Section 1.7.1.

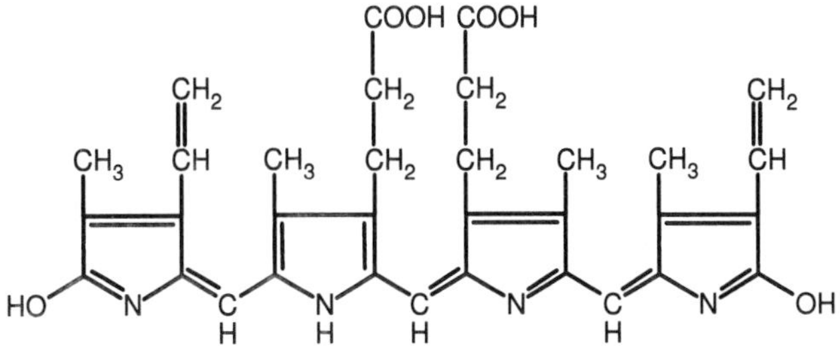

(9)

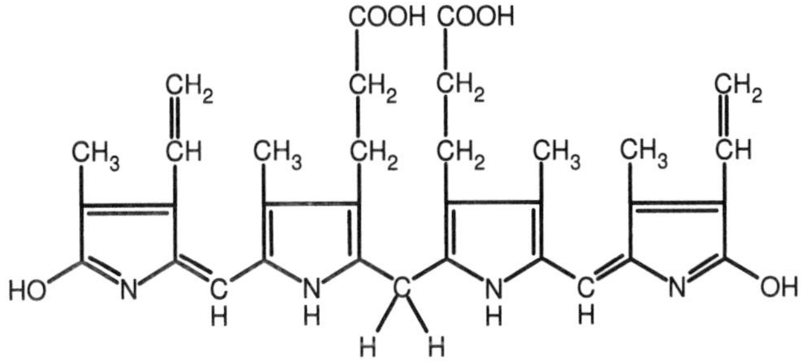

(10)

Cells of algae, cyanobacteria, and other photosynthetic organisms often contain bile pigments known as phycobilins. These compounds function both as primary and secondary photosynthetic pigments in the organisms concerned. In the functional state, they are covalently linked with specific proteins and are responsible for the characteristic colour of the organisms. The two major classes of phycobilins are known as phycocyanins (blue pigments) and phycoerythrins (red pigments). Their respective chromophores are shown in Fig. 1.4a,b. In blue-green and red algae, phycobiliproteins form functional aggregates called phycobilisomes.

Phytochrome, the light-receptor molecule for a number of important, non-photosynthetic, physiological functions in plants is also a biliprotein; its chromophore (Fig. 1.4c) closely resenbles that of the phycobilins. The compound is of ubiquitous occurrence in higher plants and is responsible for such light-dependent responses as photoperiodism and phototropism.

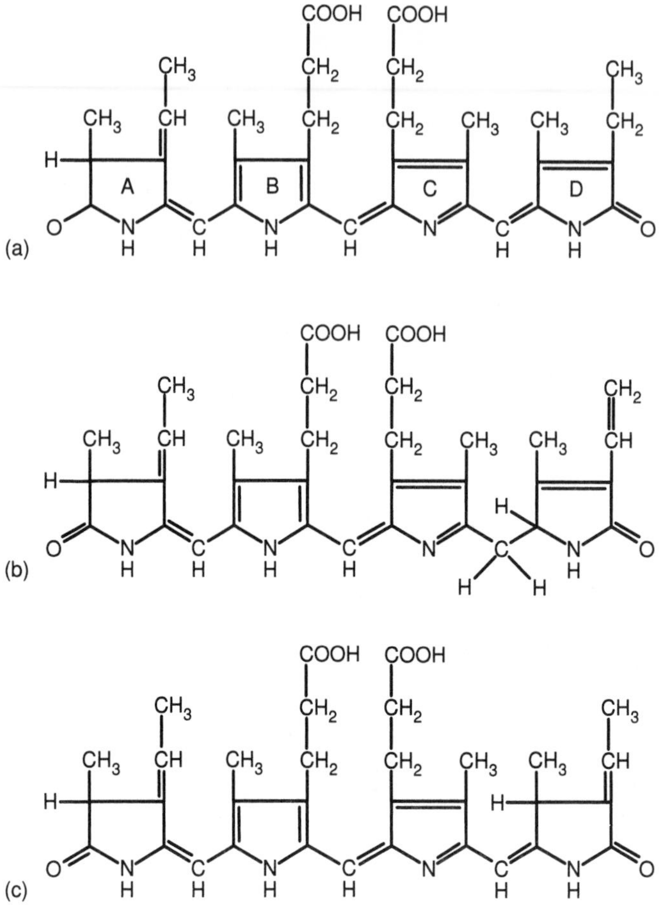

(a)

(b)

(c)

Fig. 1.4 Structure of the linear tetrapyrrole (bile pigment) chromophores of the phycobilins: (a) phycocyanin (blue) and (b) phycoerythrin (red). The structure of the related light-receptor phytochrome, which indicates light-induced physiological responses in plants, is shown in (c).

1.4.3 Corrins

The ring system of the corrins, formerly known as pseudoporphyrins, is closely related to that of the porphyrins but one of the methene bridges is missing and rings A and D are linked directly through their adjacent α-carbon atoms. This type of structure is typified by that of vitamin B_{12} (11) biochemically the most important corrin. Dietary deficiency of this vitamin results in pernicious anaemia. The complex three-dimensional structure of vitamin B_{12}, elucidated in 1956 by the notable X-ray crystallographic studies of Dorothy Hodgkin, comprises three main components. The central, planar, corrin unit, complexed with cobalt, is

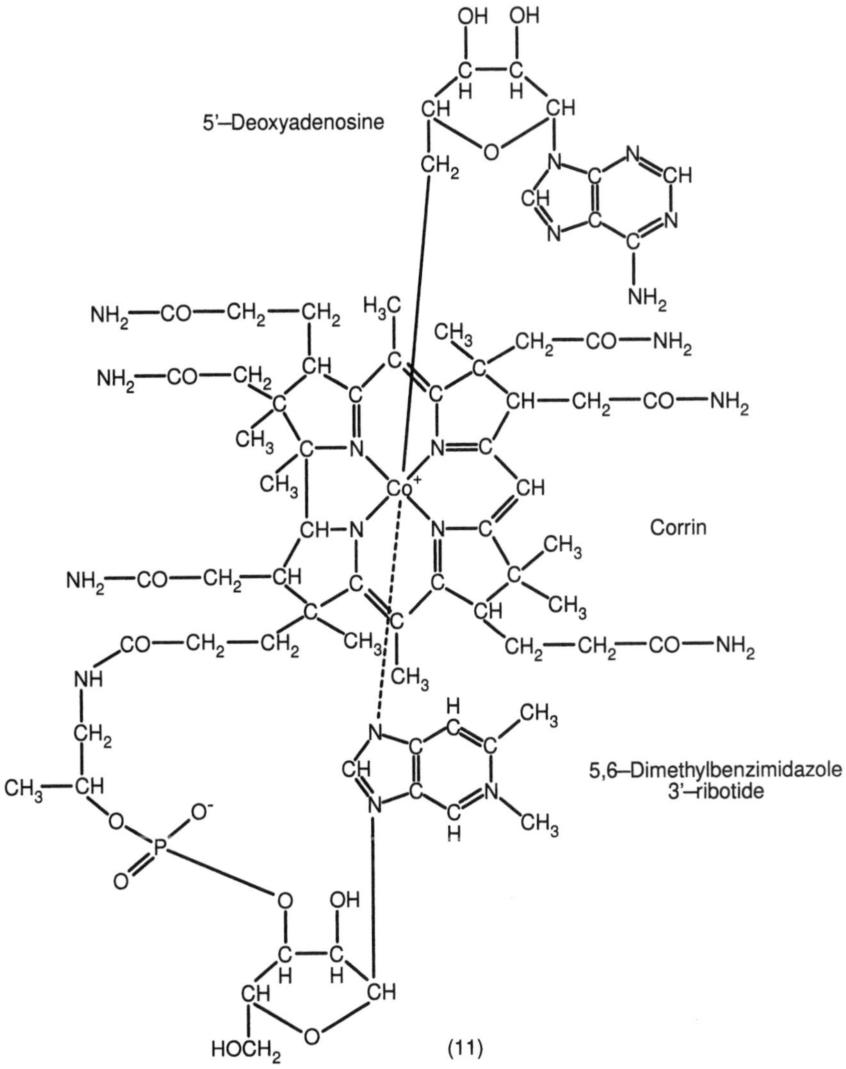

5′–Deoxyadenosine

Corrin

5,6–Dimethylbenzimidazole 3′–ribotide

(11)

attached through the Co^+ ion to a 5′-deoxyadenosine residue on one side, and on the other side to a nucleotide-like structure, 5,6-dimethylbenzimidazole 3′-ribotide. This is an analogue of adenosine 3′-monophosphate in which the adenine ring is replaced by that of benzimidazole.

1.5 NATURAL OCCURRENCE OF DIPYRROLES AND TRIPYRROLES

Both dipyrroles and tripyrroles are rare in nature and occur mainly as transient intermediates in porphyrin and bile pigment metabolism. An exception is the unusual, crimson, pigment prodigiosin, produced by the microorganism *Serratia marcescens*. This tripyrrole was originally assigned a tripyrrylmethene structure but this was later revised to a pyrryldipyrrylmethene arrangement (12).

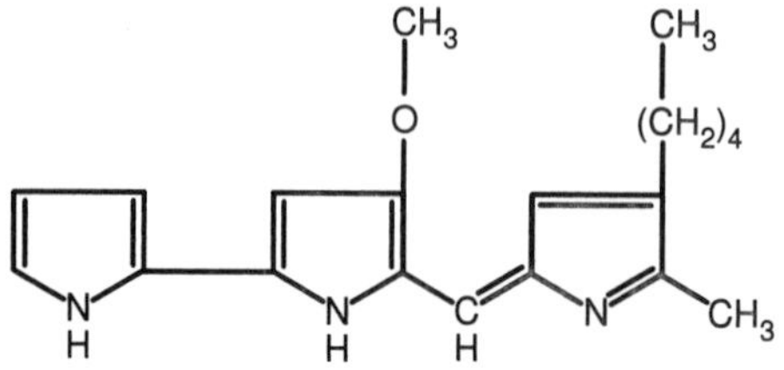

(12)

1.6 PORPHYRIN BIOSYNTHESIS AND INTERCONVERSION

1.6.1 Haem

Elucidation of the pathway of porphyrin biosynthesis began with the discovery of Shemin and Rittenberg in 1946 that ^{15}N-labelled glycine, orally administered to a human subject (Shemin himself) labelled the haem of haemoglobin. Attempts to use fresh mammalian blood for *in vitro* biosynthetic studies were unsuccessful but it was observed that blood rich in reticulocytes, from anaemic animals, was effective. The explanation is that during maturation to form erythrocytes, mammalian reticulocytes extrude their nuclei, and it is only nucleated cells that possess the complete biochemical machinery for haem formation. This is the rationale behind the subsequent use of duck and chicken erythrocytes for biosynthetic studies; mature avian and reptilian erythrocytes retain their nuclei. Interestingly, the mature human erythrocyte possesses the middle section of the biosynthetic pathway; it is the initial and final stages that are lost. The bone marrow cells, where production of erythrocytes (erythropoiesis) occurs in mature mammals, can of course carry out the whole of the porphyrin biosynthetic process but do not afford easily accessible experimental material.

Shemin and his collaborators, in a further series of studies, developed a stepwise degradation of the haem molecule to locate incorporated isotopes. They observed that all four *N*-atoms of the porphyrin ring originated from glycine and that the methylene *C*-atom of glycine contributed all four bridge *C*-atoms together with one other *C*-atom in each pyrrole ring. The carboxyl carbon of glycine was found not to be incorporated into the porphyrin structure.

More detailed examination of the labelling pattern obtained in the haem molecule from ^{14}C-acetate indicated that there is a precursor of the porphyrin ring which derives from glycine, and a second that arises from acetate *via* the tricarboxylic acid cycle. The latter was subsequently shown to be succinyl-CoA which combines with glycine to form 5-aminolaevulinic acid (13). Shemin postulated that a pyrrolic intermediate is formed during porphyrin biosynthesis but at that time its precise structure was uncertain. What proved to be a decisive discovery was made by Westall in 1952 when he isolated the pyrrole 'porphobilinogen' from the urine of a patient suffering from acute porphyria, a disease of porphyrin metabolism. This compound was shown by Cookson & Rimington (1954) to have the structure (14) and by Falk and his collaborators (1953), working in Rimington's laboratory, to be converted by chicken blood haemolysates into uroporphyrin, coproporphyrin, protoporphyrin, and haem, in that order.

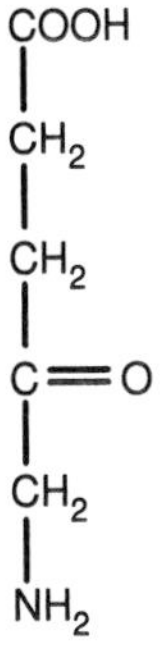

(13) 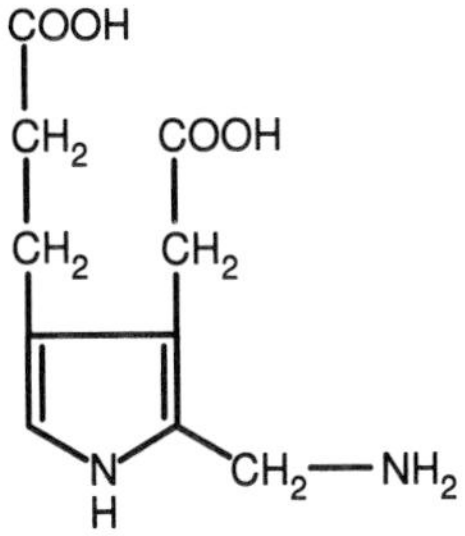 (14)

To consider present-day knowledge of the haem biosynthetic pathway, it is useful to divide it into three phases. First is formation of the pyrrolic precursor of the porphyrin macrocycle, i.e. porphobilinogen. In the initial step (Fig. 1.5) succinyl-CoA, generated by the tricarboxylic acid cycle, is condensed with glycine in a mitochondrial process catalysed by the enzyme 5-aminolaevulate synthase (ALA-synthase) which requires pyridoxal 5′-phosphate as a coenzyme. The level of activity of this enzyme constitutes a rate-limiting step in haem synthesis, and it is markedly affected by various drugs. Haem is an inhibitor of ALA-synthase and as the end-product of the biosynthetic sequence, it serves as a negative feedback control, so preventing overproduction.

The reaction sequence, shown in Fig. 1.5a, leading to synthesis of 5-aminolaevulinic acid (ALA) involves spontaneous intermediate formation of a Schiff's base between glycine and pyridoxal 5′-phosphate. Loss of a proton from the α-carbon of the Schiff's base yields a stabilized carbanion (Fig. 1.5b) which reacts with the electrophilic carbonyl-carbon of succinyl CoA to yield 2-amino-3-oxoadipic acid. Being a β-keto acid, this spontaneously decarboxylates to produce ALA. The whole of the sequence, from glycine and succinyl CoA to ALA, takes

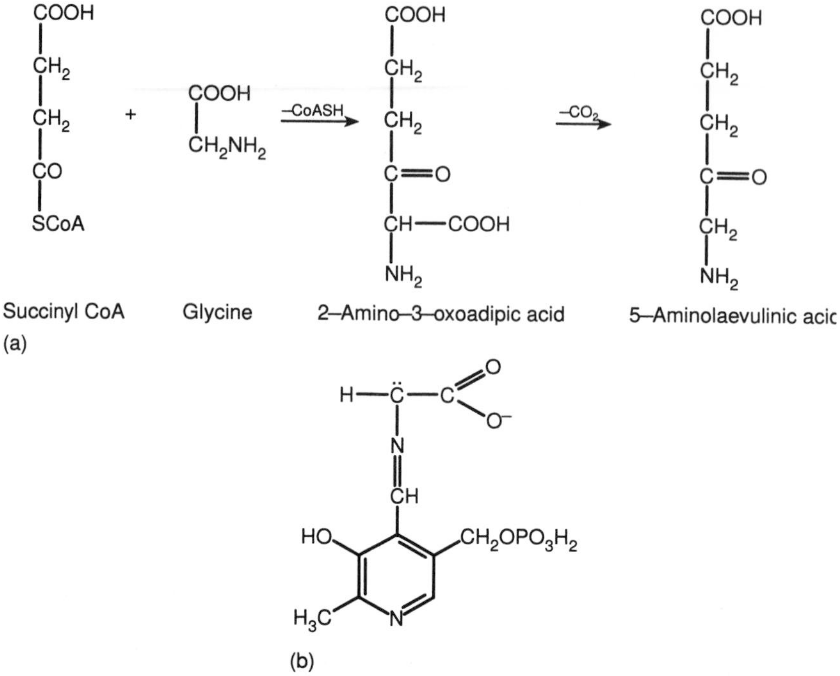

Succinyl CoA + Glycine —CoASH→ 2–Amino–3–oxoadipic acid —CO_2→ 5–Aminolaevulinic acic

(a)

(b)

Fig. 1.5 Initial stages in the biosynthesis of haem. (a) Succinyl-CoA, generated by the tricarboxylic acid cycle, is condensed with glycine to form 2-amino-3-oxoadipic acid which spontaneously decarboxylates to give 5-aminolaevulinic acid (ALA). The reaction is catalysed by the pyridoxal 5′-phosphate-dependent enzyme ALA-synthase and involves intermediate formation of a Schiff's base between glycine and pyridoxal 5′-phosphate. Loss of a proton leads to a stabilized carbanion (b) which interacts with the electrophilic carbonyl-carbon of succinyl CoA, giving 2-amino-3-oxoadipic acid.

place at the enzyme surface and the intermediates have no free existence.

The second phase of haem biosynthesis comprises formation of the pyrrolic intermediate, porphobilinogen, and the use of four of these molecules to assemble a porphyrin ring. The biosynthetic route is shown in Fig. 1.6. It initially involves condensation of two molecules of ALA, a reaction catalysed by the Zn-requiring enzyme ALA-dehydratase, to form a molecule of porphobilinogen. The currently accepted model for this enzymic reaction (Battle & Stella, 1978) involves a minimal functional dimer of the dehydratase. One molecule of ALA forms a Schiff's base with the 5-amino group of a lysine residue while the second ALA molecule is held non-covalently through its carboxyl group to a positively charged residue in the protein. The overall effect is that the two substrate molecules sit in a hole in the enzyme protein

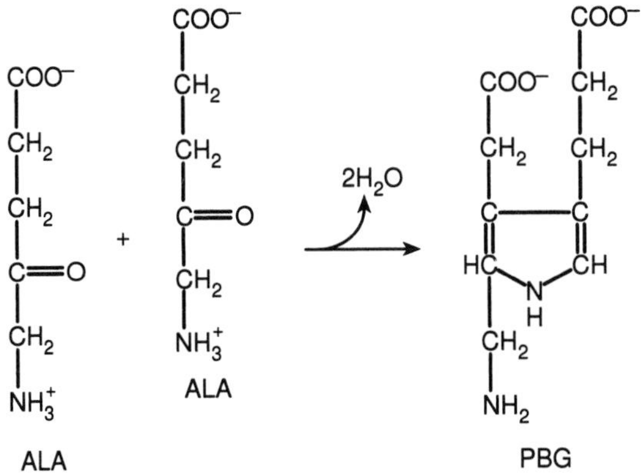

Fig. 1.6 Two molecules of ALA are condensed by the enzyme ALA-dehydratase to yield a molecule of porphobilinogen.

and neighbouring histidine residues control the uptake and release of protons upon which the formation of the porphobilinogen molecule depends.

Biological systems normally produce porphyrins of the isomeric series III but work in Rimington's laboratory (Rimington & Booij, 1957) showed that if red-cell haemolysates were pre-heated briefly to 60°C, only uroporphyrin I and coproporphyrin I were produced in subsequent incubations with porphobilinogen. Bogorad (1958) had already shown the presence in plants of two enzymes, a porphobilinogen deaminase which, by itself, catalyses formation of coproporphyrin I, and a cosynthase, inactive alone, but necessary in conjunction with the deaminase to produce coproporphyrin III. It was apparent that the heat-labile component of red-cell haemolysates is the cosynthase. Consideration of the formulae of uroporphyrin I and uroporphyrin III (Fig. 1.2) indicates that the overall effect of presence of the cosynthase is to reverse the substituents on ring D. Thus it appears that the deaminase produces a regular tetrapyrrole with positions 1–8 being occupied by acetic acid (A) and propanoic acid (P) residues, in the sequence A, P, A, P, A, P, A, P, respectively but that the cosynthase reverses the last two, to give A, P, A, P, A, P, P, A. The intriguing question is, how is this done? Battersby (1980, 1985) solved the major part of the problem, producing evidence that the deaminase catalyses formation of a regular linear tetrapyrrole and that the cosynthase then catalyses ring closure of the macrocycle to form a spirane which rearranges to give the series III isomer. Some spontaneous cyclization of the linear tetrapyrrole also occurs, yielding small amounts of the series I isomer. The current model of the process is shown in Fig. 1.7 from which it can be seen that the initial products are not, as was originally thought, uroporphyrins but

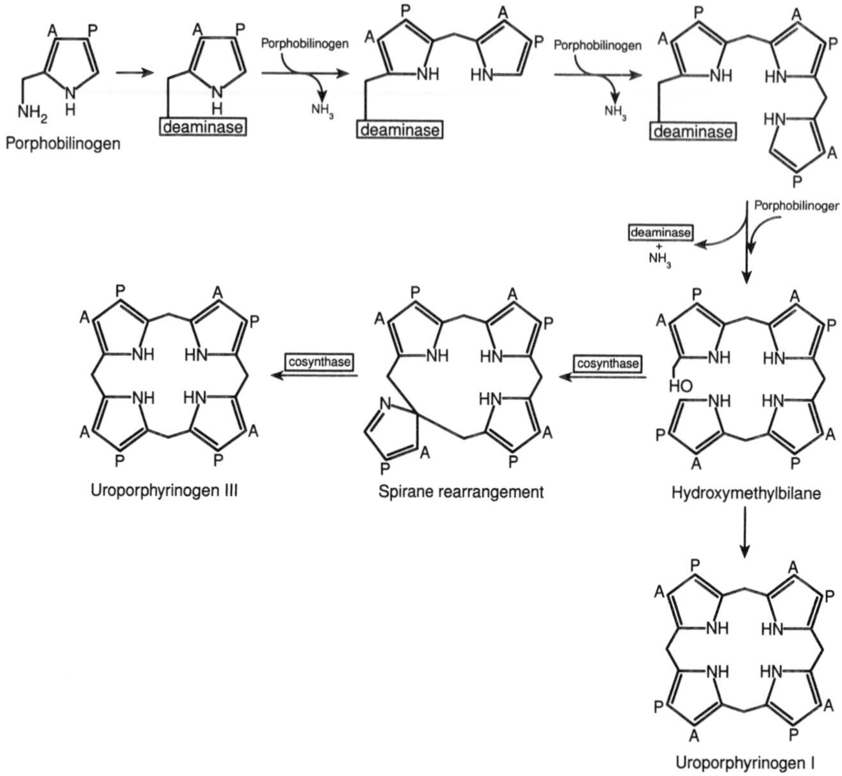

Fig. 1.7 Current model of the biosynthetic system producing uroporphyrinogens I and III from porphobilinogen (PBG). A and P represent acetic and propanoic acid residues, respectively. The enzyme PBG deaminase combines two molecules of PBG, releasing the amino group from one of them, to form an enzyme bound dipyrrole intermediate. A third PBG unit is added, then a fourth, to yield a linear tetrapyrrole hydroxymethylbilane which can ring close to form uroporphyyrinogen I. However, under the influence of a second enzyme, cosynthase, the hydroxymethylbilane ring forms an intermediate spirane which permits rotation of the fourth pyrrole ring before final closure to yield uroporphyrinogen III. In effect, the cosynthase changes the substitution pattern from that of a I series porphyrin to that of a III series compound (cf. Fig. 1.1).

uroporphyrinogens. Porphyrinogens differ from their porphyrin counterparts by having six extra H-atoms per molecule. Structurally, the porphyrinogens have four methylene bridges instead of methene bridges, and two of the pyrrolic N-atoms have been reduced (Fig. 1.8).

As with uroporphyrinogen, the biosynthetic process results in the production of coproporphyrinogen and protoporphyrinogen but these reduced *'ogen* forms are very easily oxidized, both spontaneously and enzymically, to their corresponding porphyrins, accounting for the natural occurrence of the latter.

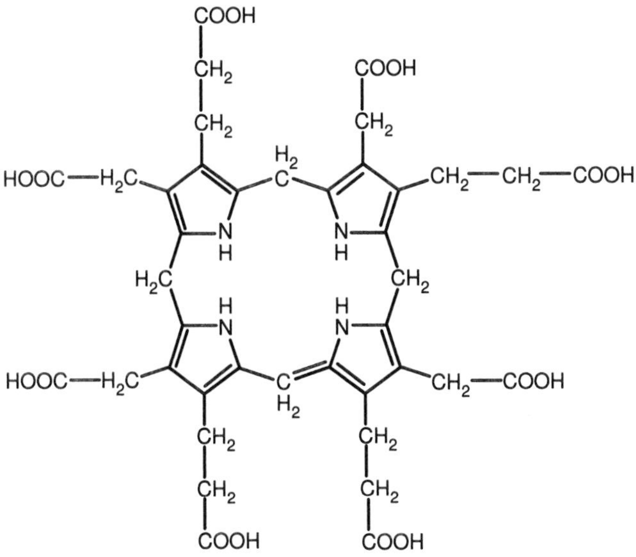

Uroporphyrinogen III

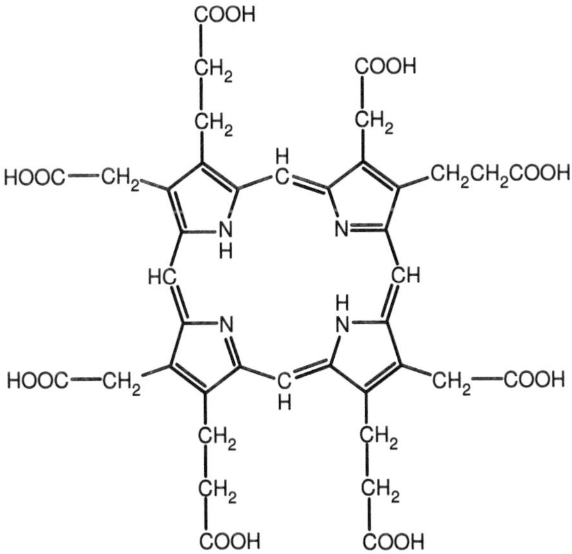

Uroporphyrin III

Fig. 1.8 Structure of uroporphyrin III and uroporphyrinogen III to illustrate the difference in structure between the porphyrins and the porphyrinogens. The latter have six extra hydrogen atoms per molecule and are the main line biosynthetic intermediates.

The final phase of haem biosynthesis is the conversion of uroporphyrinogen III to protoporphyrin IX (Fig. 1.9) and the insertion of Fe^{2+}. Clues to the mechanism by which uroporphyrinogen III is converted to protoporphyrin IX came from the discovery by Jackson, Elder and their co-workers (Jackson *et al.*, 1976) of porphyrin hepta-, hexa-, and pentacarboxylic acid species in the faeces of rats poisoned with hexachlorobenzene, a known inducer of a toxic porphyria. Since the molecule of uroporphyrinogen possesses eight carboxyl groups whereas protoporphyrin IX has only two, it is evident that the biosynthesis of haem must involve loss of 6 carboxyls. Discovery of the hepta-, hexa-, and pentacarboxylic acids indicated that there is a stepwise decarboxylation of uroporphyrinogen III taking place. Removal of each of the first four carboxyls is catalysed by the enzyme uroporphyrinogen III decarboxylase, and the process ends in formation of coproporphyrinogen III. The enzyme coproporphyrinogen III oxidase removes the remaining two carboxyl groups whilst simultaneously removing four hydrogens (Fig. 1.9). Characteristically of an oxidase, the enzyme uses atmospheric oxygen as its hydrogen acceptor. The product of this oxidative decarboxylation is protoporphyrinogen IX which is further oxidized, also enzymically, to convert it from the porphyrinogen form to protoporphyrin IX. Insertion of Fe^{2+} into this molecule produces haem (Fig. 1.9). Porphyrins are, under physiological conditions, good chelating agents and haem can be formed non-enzymically from protoporphyrin IX if Fe^{2+} ions are readily available. Nevertheless, there is an enzyme available to facilitate iron insertion. Ferrochelatase, as it is called, is not substrate specific and will catalyse chelation of different divalent metal ions by various dicarboxylic porphyrins. Probably because of its ready availability in the haem-synthesizing cells, Fe^{2+} predominates in this respect. Lipids are essential for the process but it is inhibited by oxygen. Presence of ascorbic acid (vitamin C) helps to maintain Fe^{2+} in the ferrous state for haem formation.

As was noted at the beginning of this section, mitochondria and nuclei are necessarily involved in haem biosynthesis as well as enzymes from the cytosol. The subcellular location and control of the various stages of the biosynthetic process are shown diagramatically in Fig. 1.10. The main regulatory process appears to be the negative feedback by haem on the synthesis of 5-aminolaevulinic acid and in particular on the availability of mitochondrial ALA synthase.

1.6.2 Chlorophyll

The biosynthesis of chlorophyll by green plants and most photosynthetic microorganisms follows an essential similar route to haem. Contrary to earlier expectations, however, it was shown by Beale and his collaborators in the mid 1970s that in these organisms ALA is not pro-

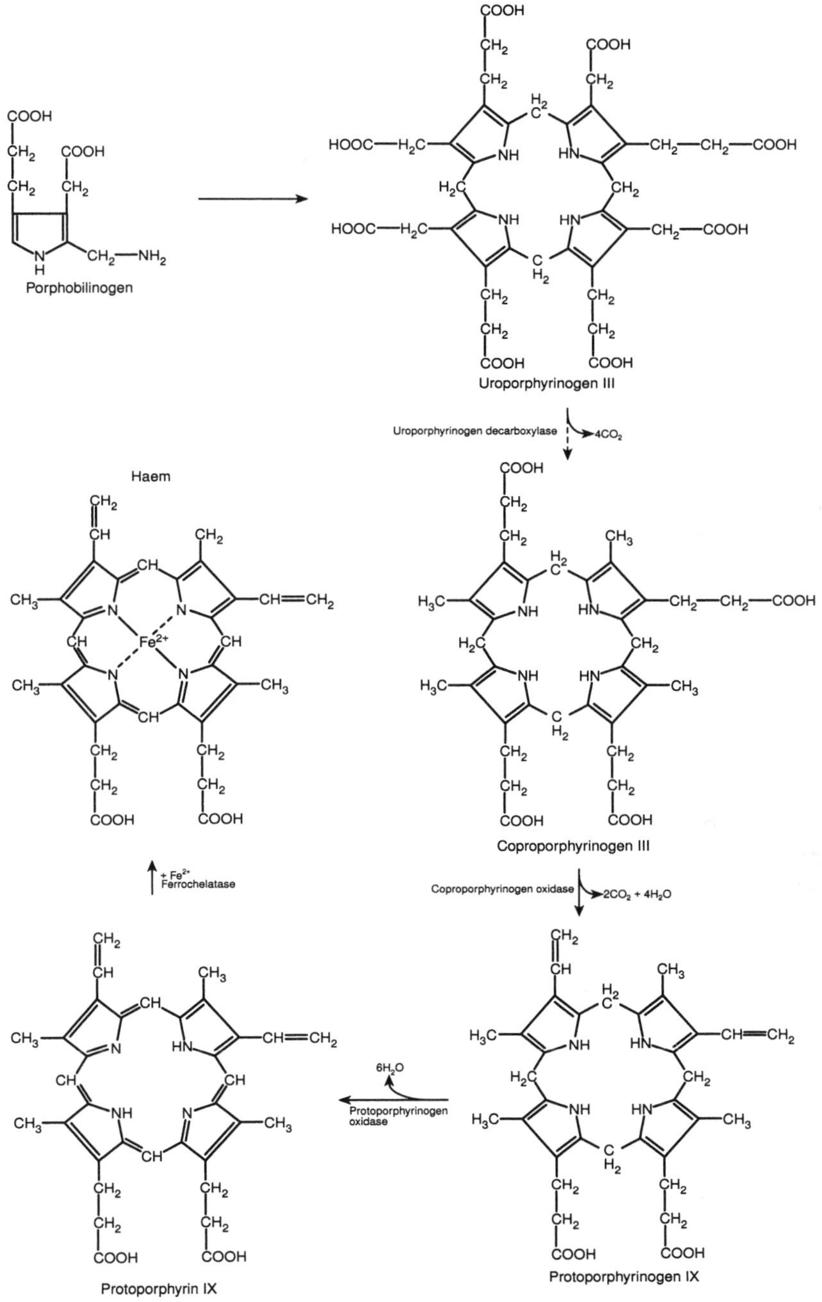

Fig. 1.9 The final phase of haem biosynthesis in which uroporphyrinogen III undergoes successive decarboxylations, enzymically, and is then oxidized to yield protoporphyrin IX. Insertion of Fe^{2+} into protoporphyrin IX to form haem is catalysed by the enzyme ferrochelatase.

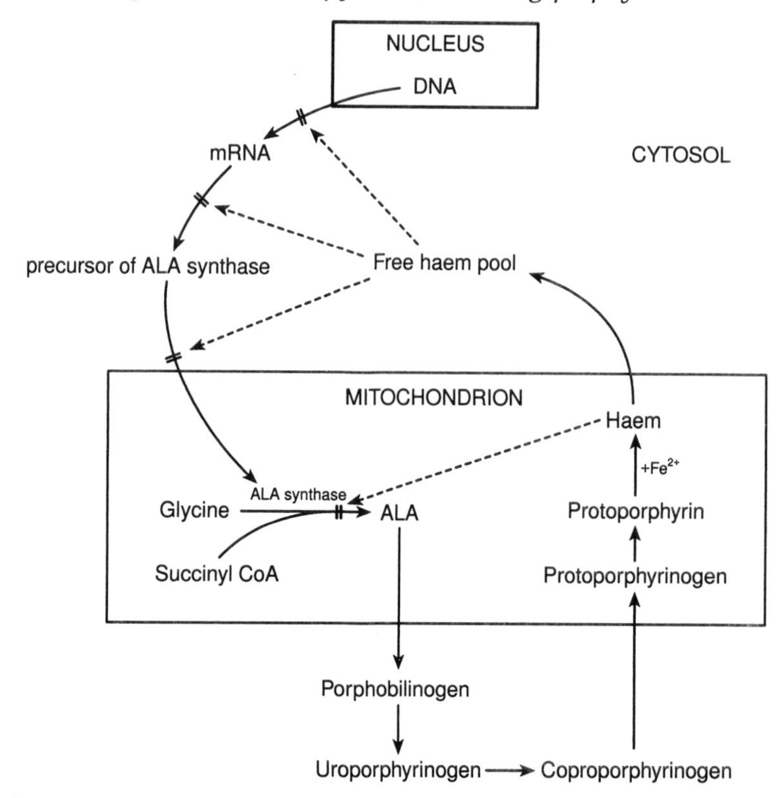

Fig. 1.10 Subcellular location and control of the various stages of haem biosynthesis. Generation of succinyl-CoA and its condensation with glycine, leading to formation of ALA, occurs in the mitochondrion. Free haem acts as a negative feedback control on this process. Oxidation of protoporphyrin IX to protoporphyrin and its conversion to haem also occurs in the mitochondrion but the other processes, also subject to feedback control by free haem, take place in the cytosol. The nucleus controls the overall mechanism through production of the necessary mRNA for ALA synthase.

duced from succinyl CoA and glycine but from glutamate, with C-1 of this amino acid giving rise to C-5 of 5-aminolaevulinic acid. The glutamate pathway of ALA synthesis, shown in Fig. 1.11, begins with ATP-requiring initial steps of protein biosynthesis in which an amino acid, in this case glutamate, forms an aminoacyl link with a specific transfer RNA molecule (tRNA). This 'activated' glutamate is reduced enzymically, by NADPH, to the semialdehyde and the tRNA molecule released. Then, in effect, the amino group migrates to the adjacent terminal C-atom, which has how become C-5, to give 5-aminolaevulinic acid. Whether this is an intra- or intermolecular amino transfer is still not clear although available evidence favours the latter. The step is catalysed by a pyridoxal phosphate-dependent aminotransferase.

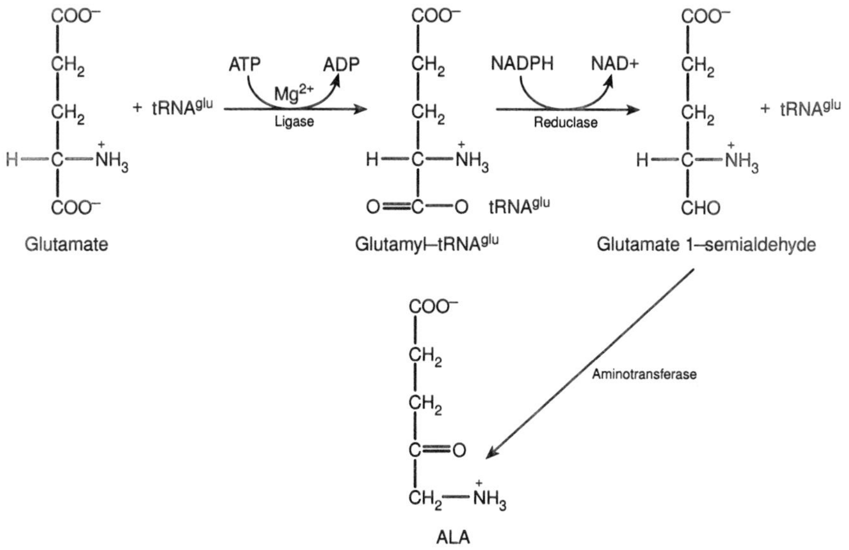

Fig. 1.11 The glutamate pathway for ALA synthesis. In higher plants, ALA originates from glutamic acid and not from glycine as in animals. The available evidence indicates enzymic reduction of glutamyl-tRNA to glutamic semialdehyde, with the release of the tRNA, and the transamination of the semialdehyde to ALA.

From ALA to protoporphyrin IX, the biosynthetic sequence follows the same pathway as that of haem formation. Green plants and photosynthetic microorganisms, of course, still need to make haem for their haem-enzymes and cytochromes, and can enzymically insert Fe^{2+} into protoporphyrin IX for this purpose. There is, however, a branch in the biosynthetic pathway at this point in photosynthetic organisms and instead of Fe^{2+}, Mg^{2+} can be inserted. The process is catalysed by a Mg chelatase. The further elaboration of Mg-protoporphyrin IX to chlorophyll is outlined in Fig. 1.12. The first step is enzymic methylation of the propionic acid residue at position 6, with S-adenosylmethionine as the methyl donor, to yield Mg-protoporphyrin IX monomethyl ester. Next, in this modified propionic acid side chain, the C-atom adjacent to the ring (C) is oxidized to an oxo group. It has been suggested that presence of the methyl group prevents the spontaneous decarboxylation of what would otherwise be a β-keto acid, and that the resulting conditions facilitate the reaction of the next C-atom in the side chain with the methene bridge, to form the isocyclic ring of the chlorophyll molecule (Fig. 1.12). The product, with the additional ring, is known as a pheoporphyrin. Specifically, it is Mg-2,4-divinylpheoporphyrin a_5 monomethyl ester. The term a_5 often puzzles newcomers to the field of porphyrin chemistry; it simply indicates that it is structurally related to

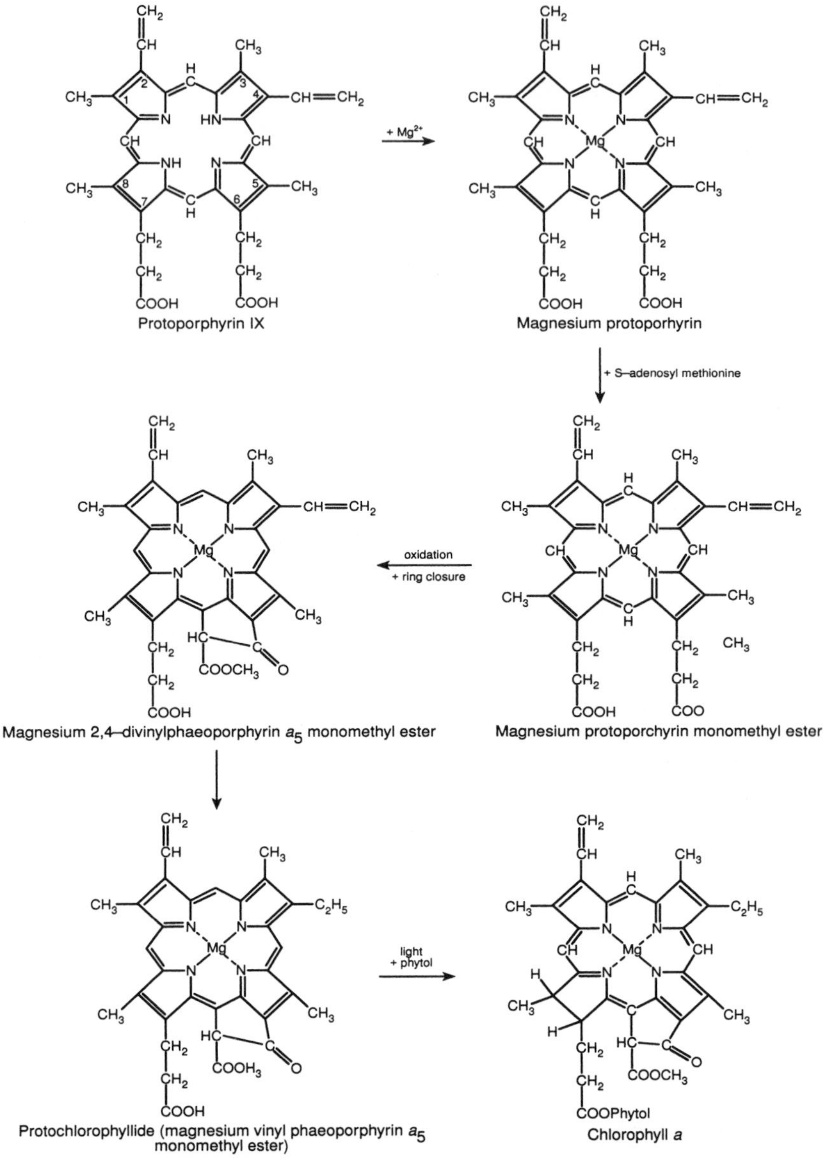

Protoporphyrin IX

+ Mg^{2+}

Magnesium protoporhyrin

+ S–adenosyl methionine

Magnesium protoporchyrin monomethyl ester

oxidation
+ ring closure

Magnesium 2,4–divinylphaeoporphyrin a_5 monomethyl ester

Protochlorophyllide (magnesium vinyl phaeoporphyrin a_5 monomethyl ester)

light
+ phytol

Chlorophyll *a*

Fig. 1.12 The biosynthesis of chlorophyll *a* from protoporphyrin IX. Insertion of Mg^{2+} into protoporphyrin IX is catalysed by the enzyme Mg-chelatase and the product is enzymically methylated, with *S*-adenosylmethionine as the methyl donor. A new isocyclic ring is then formed on ring C by oxidation of the adjacent side chain and ring closure. The new phaeoporphyrin is enzymically reduced to protochlorophyllide by reduction of the vinyl at ring B to an ethyl group. The final step shown, from protochlorophyllide to chlorophyll *a*, is in fact two sequential reactions. First protochlorophyllide is reduced to chlorophyll *a* in a light-dependent reduction, and then chlorophyllide *a* is esterified with phytol to yield chlorophyll *a*.

chlorophyll *a* and that it possesses 5-oxygen atoms per molecule. In the penultimate steps of chlorophyll biosynthesis, the 4-vinyl group of the pheoporphyrin is enzymically reduced to ethyl and the product, known as protochlorophyllide, is further reduced in a light-dependent reaction in which ring D acquires two *H*-atoms *trans* to one another. Chlorophyllide *a*, so formed, is finally converted to chlorophyll *a* by esterification with the long chain terpenoid alcohol phytol (8). This esterification occurs at the remaining propionic acid residue, on ring D.

Two points of particular biochemical relevance in the foregoing biosynthetic process are worthy of attention. One is the photoreduction of protochlorophyllide. This process explains the light-requirement for chlorophyll formation; plants grown in the dark become 'etiolated' and almost devoid of chlorophyll. Second, is the effect of esterification with phytol. Whereas the main part of the chlorophyll molecule is hydrophilic, the phytyl side chain is lipophilic. This modication thus facilitates the functional disposition of the chlorophyll molecule in the chloroplast membrane, with one end of the molecule in the lipid phase and the other in the aqueous phase.

Chlorophyll *b* only differs structurally from chlorophyll *a* in having a formyl group instead of a methyl at position 3 (Fig. 1.3) and it has been generally accepted for some time that chlorophyll *b* is formed by oxidation of chlorophyll *a*. There is, however, some published work suggesting that the oxidation may take place before phytylation, i.e. the biosynthetic sequence may be chlorophyllide *a* to chlorophyllide *b* to chlorophyll *b*. Bacteriochlorophyll (Fig. 1.3) differs from chlorophyll *a* by having an acetyl- instead of a vinyl group at position 2, and ring B is reduced. Little is known of the details of the formation of this compound but it involves an essentially similar biosynthetic route to that by which the chlorophylls are produced in higher plants.

1.7 PORPHYRIN CATABOLISM AND THE FORMATION OF BILE PIGMENTS AND PHYCOBILINS

1.7.1 Bile pigments

The normal human erythrocyte has a life span of about 120 days. Older, effete, cells are removed from circulation by the reticuloendothelial system, comprising the spleen and the phagocytic cells of the blood. Lysis of red cells during this disposal process releases haemoglobin which is catabolized to form bile pigments. In certain pathological conditions, significant numbers of erythrocytes can undergo spontaneous haemolysis, releasing abnormally large amounts of haemoglobin into the blood plasma. This excessive loss of haemoglobin can not only result in 'haemolytic anaemia' but by embarrasing the catabolic process, it can also cause accumulation of bile pigments, particularly in the skin and conjunctiva, giving rise to the characteristic yellowing known as jaundice.

The initial catabolic steps through which haemoglobin passes, take place in the reticuloendothelial cells (Fig. 1.13). Haemoglobin is first hydroxylated at the methene bridge between rings A and B, and this results in loss of the methene C-atom as carbon monoxide. This is a somewhat surprising biological product in view of the toxicity attributable to its tendency to bind with haem. The poisonous effect is, however, minimized in this context by the three-dimensional shape of the oxygen-binding site of haemoglobin and myoglobin, which discriminates against CO and in favour of oxygen. Furthermore, oxygen is usually present in normal tissues at much higher concentrations than CO. It has been calculated that fewer than 1% of the total oxygen-binding sites of haemoglobin and myoglobin are blocked at any one time by CO in normal healthy tissues.

Verdoglobin, the product of the opening of the haemoglobin macrocycle (Fig. 1.13) is a dark green pigment. It is this that is responsible for the characteristic green tinge accompanying bruises and black eyes. Erythrocytes in the skin capillaries, damaged by a blow or sustained pressure, rapidly haemolyse and release haemoglobin which begins to be catabolized *in situ* by the circulating reticuloendothelial cells of the blood. Further, hydrolytic, degradation of verdoglobin yields Fe^{3+} which is transported by a special carrier protein, to be reused in erythropoiesis and other iron-requiring processes. The released globin component of haemoglobin undergoes enzymic hydrolysis to its constituent amino acids, and these are also salvaged.

At this stage in haemoglobin catabolism, the linear tetrapyrrole residue is biliverdin (Fig. 1.13) the main green component of bile. Enzymic reduction of the central methene bridge of this compound, to form a methylene bridge, yields bilirubin, the major red constituent of bile (Fig. 1.13). Bilirubin is a relatively insoluble compound and in order to be transported to the liver where it can be disposed of in the bile, it needs to be solubilized by conjugation with albumin. Failure to conjugate bilirubin in this way leads to an accumulation of the pigment, resulting in jaundice. Blockage of the bile duct, e.g. by a gall stone, prevents the bile pigments from reaching the gut, where the final stages of bile pigment metabolism normally occur, and this condition also results in jaundice.

The intestinal flora metabolizes bilirubin, initially by reduction to mesobilirubinogen and then by further reduction to urobilin IX_{α} and stercobilin (Fig. 1.14). Some reabsorbtion of these reduced products occurs and they are transported in this way, through the bloodstream, to the kidneys where they are excreted in the urine, giving it its familiar yellow colour.

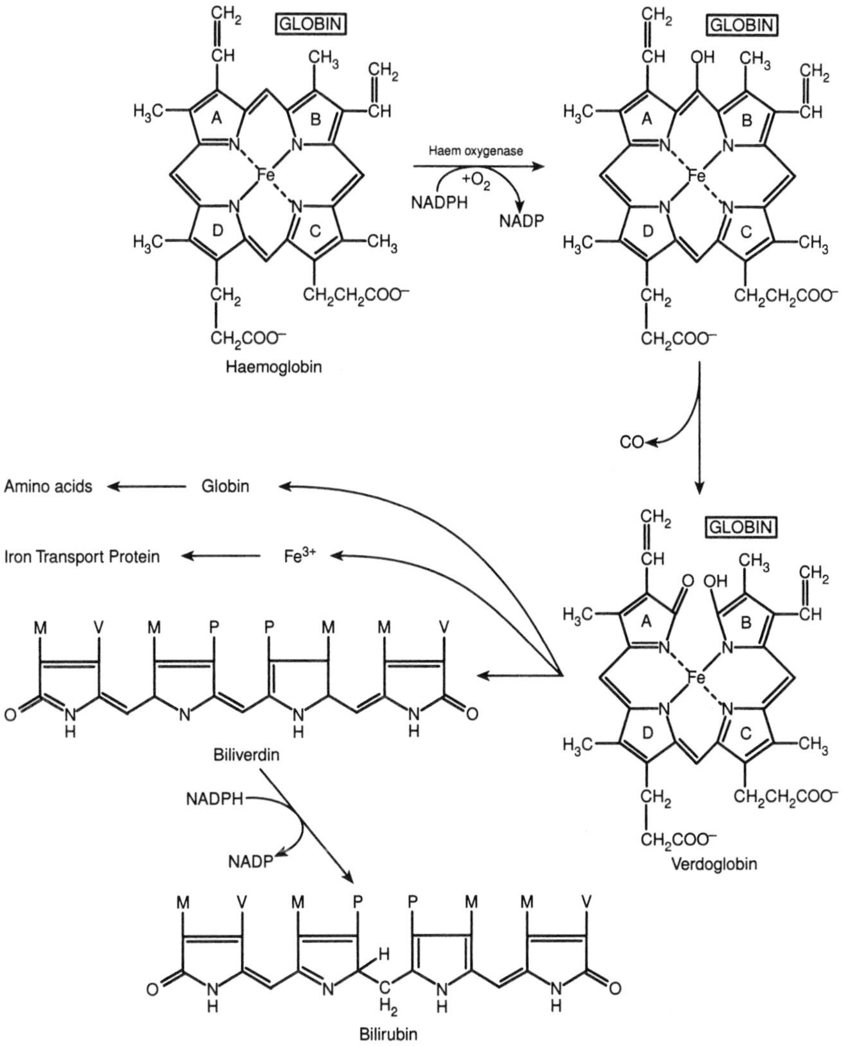

Fig. 1.13 Catabolism of haemoglobin and the formation of the bile pigments biliverdin and bilirubin. Enzymic oxidation of the methene bridge between pyrrole rings A and B of haemoglobin yields a hydroxy derivative. After ring-opening with loss of the methene carbon atom as carbon monoxide, verdochrome is formed. This is the green pigment characteristic of bruises. Next the protein component, globin, is removed together with the iron atom. Globin is hydrolysed to its component amino acids and the iron is salvaged by a special iron-transport protein. The end-product of the overall catabolic process is the linear tetrapyrrole biliverdin which is reduced to form bilirubin. Biliverdin and bilirubin are the main green and red constituents, respectively, of bile.

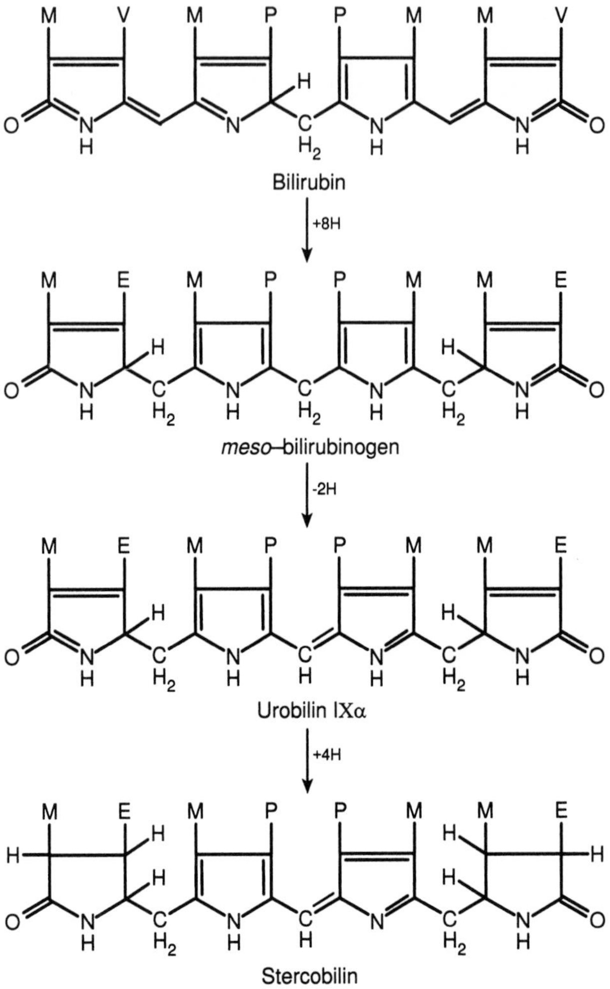

Fig. 1.14 Further metabolism of bilirubin by the intestinal flora. As bilirubin passes down the intestinal tract, it is subjected to microbial action. This mainly involves reduction and leads to formation of urobilin and stercobilin some of which is reabsorbed and excreted by the kidneys giving urine its characteristic colour.

1.7.2 Formation of phycobilins

Evidence that in the photosynthetic organisms that produce them phycobilins are formed from haem comes from two observations. ^{14}C-Haem has been shown to contribute label to phycocyanobilin in greening *Cyanidium* cells whereas chlorophyll synthesized at the same time was not labelled. Secondly, non-radioactive haem was found to decrease the incorporation of ^{14}C-ALA into phycocyanobilin. The indications are that haem is degraded to biliverdin in a similar process to that described in Section (iii) for bile pigment formation, and that the biliverdin formed is then reduced to phycocyanobilin (15) which is the precursor of the other phycobilins, such as the red pigment phycoerythrobilin (16).

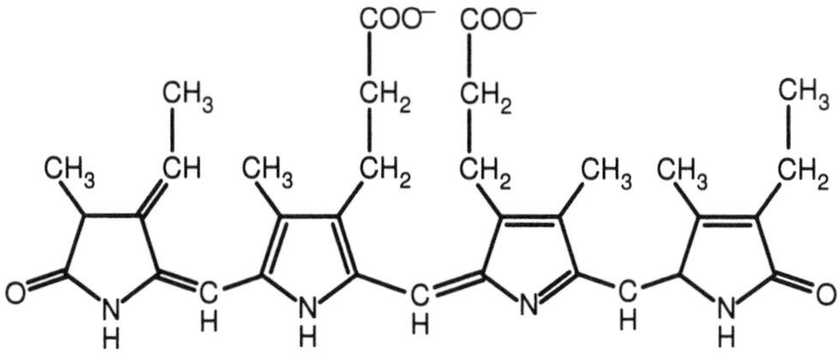

(15)

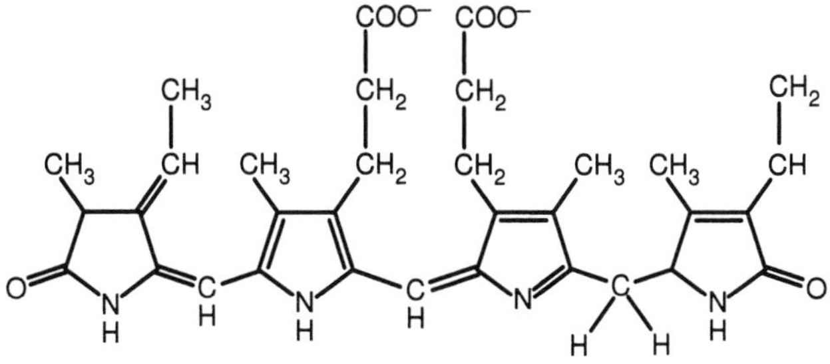

(16)

1.7.3 Formation of phytochrome

Phytochrome, the light receptor in a number of non-photosynthetic light-responses in plants, is a biliprotein of which the chromophore (17) is structurally very similar to that of the phycobilins. Evidence has been presented that, like phycocyanobilin, phytochrome can be formed *in vivo* from exogenously supplied biliverdin. This has been confirmed by use of ^{14}C-labelled biliverdin.

1.8 BIOCHEMICAL FUNCTIONS

Of the pyrrole derivatives discussed in this Chapter, the most important from a biochemical viewpoint are the haems and chlorophylls. They are the major pigments of living systems and essential catalysts in aerobic respiration and photosynthesis, respectively.

1.8.1 Haemoglobins

Lemberg and Legge (1949), in their classical treatise on haems and bile pigments, defined haemoglobins as a class of ferrous (Fe^{2+})-porphyrin proteins able to combine reversibly with oxygen without oxidation of the iron to the ferric (Fe^{3+}) state. In multicellular, highly differentiated organisms that use oxygen as a terminal electron-acceptor, haemoglobin functions as an oxygen-transporter. Oxygen is carried in this way, by the erythrocytes, from gaseous exchange sites in lungs or gills to tissues not directly in contact with the environment. Oxygen-loaded haemoglobin is known as *oxyhaemoglobin* and when unloaded it becomes *deoxyhaemoglobin*. Haemoglobin also transports, in a reverse direction, H^+-ions and CO_2.

The binding and release of transported oxygen is a complex physicochemical process dependent on three types of allosteric effect. First, the binding of oxygen is cooperative, i.e. within the same haemoglobin molecule the binding of oxygen to one haem site causes conformational changes in the protein as a whole which facilitate the binding of oxygen to the other haem sites. There are four such sites per haemoglobin molecule, each associated with a separate polypeptide subunit. Second, presence of H^+-ions and CO_2 promotes the release of transported oxygen from haemoglobin whereas presence of oxygen promotes the release of transported H^+-ions and CO_2. Thus, H^+-ions and CO_2 are carried away from the metabolic sites to the lungs, and oxygen is transported in the reverse direction. The third allosteric effect is regulation by 2,3-bisphosphoglycerate of the affinity of haemoglobin for oxygen. 2,3-Bisphosphoglycerate is a product of oxidative tissue metabolism and it binds tightly to deoxyhaemoglobin but not to oxyhaemoglobin. In consequence, its presence lowers the oxygen affinity of haemoglobin by a factor of 26. This is an essential part of the mechanism enabling haemoglobin to unload oxygen in tissue capillaries.

Myoglobin, a related haem protein involved in oxygen transport and storage, functions in a similar way to haemoglobin but its activities are confined to muscle tissue. Haem proteins with properties like those of haemoglobin have also been detected spectroscopically in a number of moulds and protozoans but their function is obscure. For many years, another biochemical paradox was the occurrence of a species of haemoglobin, known as leghaemoglobin, in the root nodules of leguminous plants. These nodules house colonies of N_2-fixing bacteria which make a major contribution to the nitrogen-economy of the plant. Plants need a plentiful supply of nitrogenous compounds to meet the requirements of protein and nucleic acid synthesis as well as production of the essential nitrogen-heterocycles with which this book is concerned. In agriculture, this need has frequently to be met by the application of expensive nitrogenous fertilizers, but leguminous plants such as peas and beans can 'fix' atmospheric N_2 into soluble compounds. Paradoxically, the nitrogen-fixing enzyme nitrogenase is strongly inhibited by oxygen yet the N_2-fixing bacteria require oxygen for their own metabolic processes. It has been calculated that 99.9% of the total oxygen in a nodule is bound to leghaemoglobin, and this, together with measurements of the kinetics of oxygen-binding to leghaemoglobin, strongly points to it being involved in the facilitated diffusion of oxygen through the nodule. It thus appears that leghaemoglobin serves to make O_2 available to the nitrogen-fixing bacteria without compromising the activity of their oxygen-sensitive nitrogenase.

1.8.2 Cytochromes

The cytochromes are a group of haem proteins found primarily in the membranes of mitochondria, chloroplasts, and the endoplasmic reticulum. They function as redox catalysts and are components of the respiratory and photosynthetic electron transport chains. Unlike haemoglobin and myoglobin, the cytochromes undergo reversible oxidation-reduction at the iron insert ($Fe^{3+} \leftrightarrow Fe^{2+}$) and during their electron transport role continually shuttle between these two states.

On the basis of their spectral absorption peaks, cytochromes are classified into three main types designated *a*, *b* and *c*. Cytochromes *b*, *c* and c_1 all have iron complexed with photoporphyrin IX. In cytochromes *c* and c_1, but not *b*, the haem is covalently linked to the protein through thioether bonds, formed between two of the vinyl side chains of haem and two of the cysteine residues of the protein. In cytochromes *a* and a_3, two of the haem side chains are modified, the vinyl at position 2 to a long (17C) hydrophobic tail, and the methyl at position 8 to formyl. This modified haem is known has haem A (18). Cytochromes *a* and a_3 each have a copper ion located close to the haem iron, and, like the haem iron, the copper undergoes reversible oxidation-reduction. Because of its relatively small size (M_r 13 000) and its solubility in water, much more is known of the structure and interactions of cytochrome *c* than of any other electron-transporting protein. Its protein component consists of a single polypeptide chain of 104 amino acid residues.

The functional sequence of the cytochromes, as electron carriers, in the electron transport chain of mitochondrial membranes is shown in Fig. 1.15. Electrons flow down the chain from the respiratory substrate to atmospheric O_2 resulting in the formation of water and the generation of ATP (see Section 6.6). When succinate is the respiratory substrate, electrons enter the chain at the ubiquinone level.

Reduction-oxidation of a cytochrome molecule involves a single electron whereas that of preceding carriers in the electron transport chain (NAD^+, FMN, FAD and ubiquinone) requires two electrons per mole-

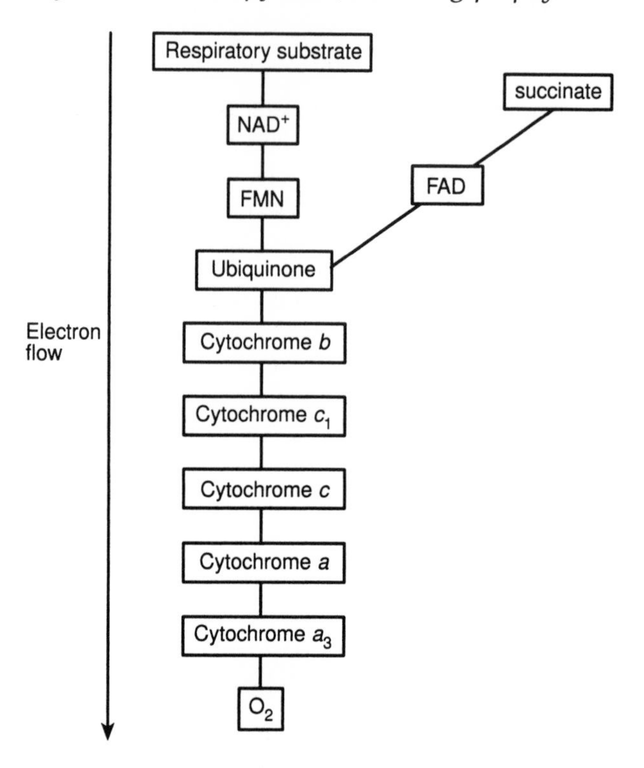

Fig. 1.15 The sequence of the cytochromes in the mitochondrial electron transport chain. Electrons flow from the oxidation of the respiratory substrate *via* NAD^+, FMN and ubiquinone to the cytochrome chain of carriers, finally being accepted by atmospheric oxygen and resulting, in the presence of a proton pool, in the formation of H_2O. Unlike the other mitochondrial dehydrogenases, succinate dehydrogenase has an FAD prosthetic group and passes electrons not to NAD^+ but directly to ubiquinone.

cule. As can be seen in Fig. 1.15, the interface between the 2-electron carriers and the single electron carriers lies between ubiquinone and cytochrome *b*. Ubiquinone seems to be particularly well suited to this interfacing role since, although requiring two electrons per molecule for full reduction, it passes through 2 single electron steps to get there, i.e. it forms a semiquinone as an intermediate (Fig. 1.16).

The enzymes peroxidase and catalase are two other biologically important haem proteins. They function in mechanisms that detoxify the biologically hazardous superoxide anion (O_2^-) formed by the partial (single electron) reduction of O_2:

$$O_2 + e^- \leftrightarrow O_2^-$$

Superoxide anions are scavenged in biological systems by the enzyme superoxide dismutase which catalyses the reaction

$$O_2^- + O_2^- \rightarrow H_2O_2 + O_2$$

However, the hydrogen peroxide formed is also potentially hazardous and has to be removed. This is effected by the respective catalytic properties of catalase and peroxidase, especially the former. Catalase is present in all aerobic organisms, in which it catalyses the decomposition of hydrogen peroxide into water and O_2, and has the distinction of being one of the most efficient biological catalysts known. It has been calculated that 1 molecule of catalase decomposes 42,000 molecules of H_2O_2 per second at 0°C. Peroxidase differs from catalase in that it

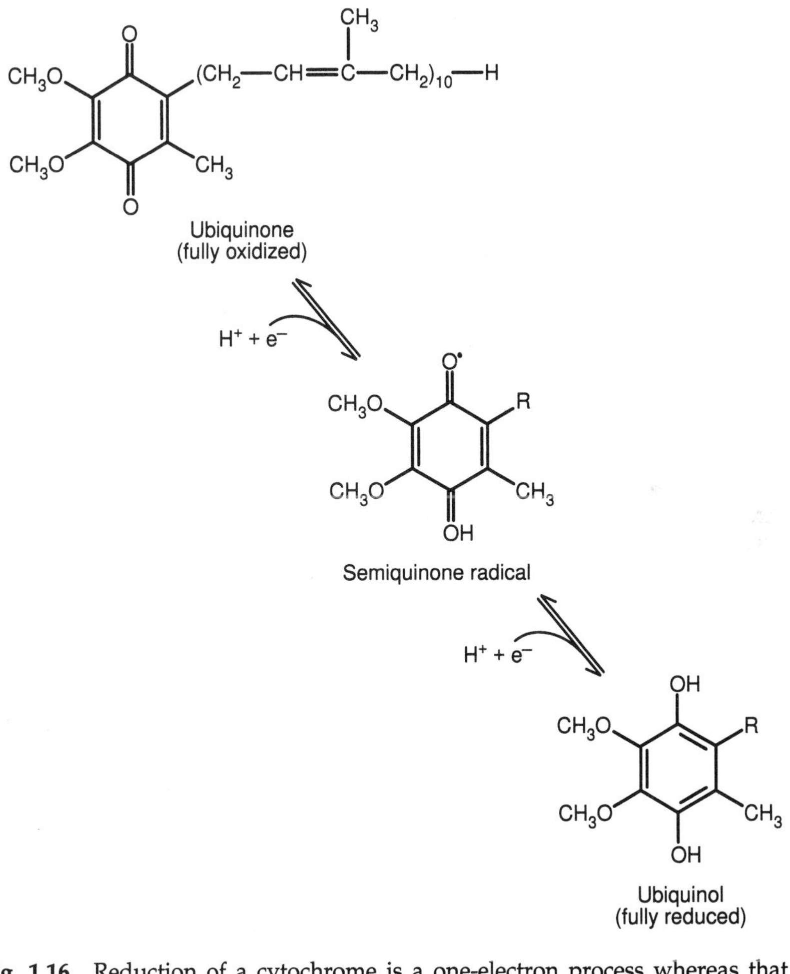

Fig. 1.16 Reduction of a cytochrome is a one-electron process whereas that of the other carriers each involves two electrons. Ubiquinone interfaces these two types of carrier in the mitochondrial electron transport chain by accepting two electrons in two single electron reductions involving formation of a semiquinone intermediate.

requires the presence of any one of a wide variety of oxidizable organic compounds (e.g. ascorbic acid) as cosubstrate (AH_2) for its activity:

$$H_2O_2 + AH_2 \rightarrow A + 2H_2O$$

Some plant tissues contain bizarrely high concentrations of peroxidase; a well known example is the root of horseradish (*Armoracia rusticana*).

1.8.3 Chlorophylls

In the tissues of green plants, the chlorophylls are located within small, highly structured, organelles called chloroplasts. Typically, these are biconvex discs with a diameter of 3–10 μm. Inside each is a series of closely-flattened sacs termed 'thylakoids' and it is the membranes of these structures that contain the photosynthetic apparatus and its associated pigments.

The photosynthetic apparatus is comprised of two distinct types of reaction centres, designated 'photosystems I and II'. These are the light-harvesting complexes. Photosystem I is a chlorophyll α-protein complex, absorbing light maximally at around 700 nm. It receives electrons from a cytochrome carrier, via a plastoquinone-protein complex, and donates electrons to the reduction of $NADP^+$. Photosystem II absorbs light maximally at a shorter wavelength of around 680 nm and although, like photosystem I, its main component is also a chlorophyll α-protein complex, it fulfills a different function. This is catalysis of the oxidation of water to oxygen and simultaneous donation of electrons, via a plastoquinone pool, to photosystem I. Most of the other photosynthetic pigments, including other chlorophylls and the carotenoids and phycobilins, are also in pigment-protein complexes associated with photosystems I and II. These 'accessory pigments' have light absorption maxima different from those of the two photosystems and they function to extend the light-harvesting wavelength range. In effect, they serve as antennae, collecting light energy, but the excitation energy is transferred, by resonance, randomly from one antenna molecule to another until it reaches a reaction centre. It has been calculated that only one O_2 is evolved for every 2500 chlorophyll molecules, indicating that most chlorophyll molecules are not directly involved in the photochemical reaction but act, as described, as antennae.

1.8.4 Phycobilins and phytochrome

The phycobilins are commonly divided into two groups on the basis of their respective colours; the red pigments are known as phycoerythrins, and the blue ones as phycocyanins. Found mainly in algae, especially in the Rhodophyceae, Cyanophyceae and Cryptophyceae, the phycobilins function in association with specific proteins as photosynthetic accessory pigments (see Section III). Whereas the chlorophylls absorb in the blue (~450 nm) and red (650–700 nm) regions of the visible spectrum, the phycobilins absorb maximally in the range 500–650 nm.

Plants show a number of morphogenic responses to light, often at intensities insignificant for photosynthesis. Examples include phototropism, release of seed dormancy and germination, induction of flowering, and anthocyanin synthesis. The main light receptor for these changes is phytochrome, a phycobilin related to but not identical with phycocyanin. Like the phycobilins in general, it is normally bound to a specific protein. In the case of the phytochrome light-receptor conjugate, the chromophore (17) is linked through a thioether bond to a cysteine residue on the protein component. Many of the phytochrome-controlled responses of plant tissues to light are induced by red light and reversed by far-red light. This is because phytochrome exists in two forms, designated P_r and P_{fr} (Fig. 1.17). The P_r form has a λ_{max} of 660 nm, and that of the P_{fr} form is at 730 nm. Irradiation of P_r with red light at around 660 nm converts it to P_{fr}; conversely, far red light at about 730 nm reconverts P_{fr} to P_r. In the dark, P_{fr} spontaneously but slowly, reverts to P_r. These relationships are outlined in Fig. 1.17.

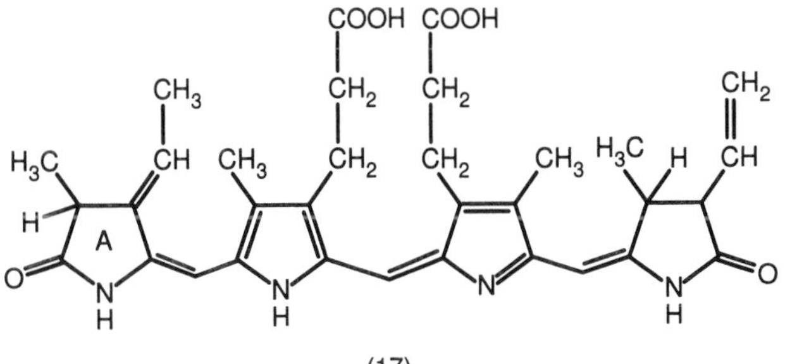

(17)

Although some details of the phytochrome-mediated light responses remain to be elucidated, it is now apparent that the P_{fr} and P_r receptors mostly function by modulating, either positively or negatively, the expression of a number of nuclear genes. There are, however, some phytochrome-mediated physiological responses in which the chromophore appears to be involved more directly, modifying the activity of a specific protein kinase or phosphoprotein phosphatase.

1.8.5 Corrins

In biological systems, the corrin ring structure is represented by vitamin B_{12} (11) and its derivatives (Fig. 1.18). Vitamin B_{12} was discovered as a result of studies of the formerly incurable disease, pernicious anaemia.

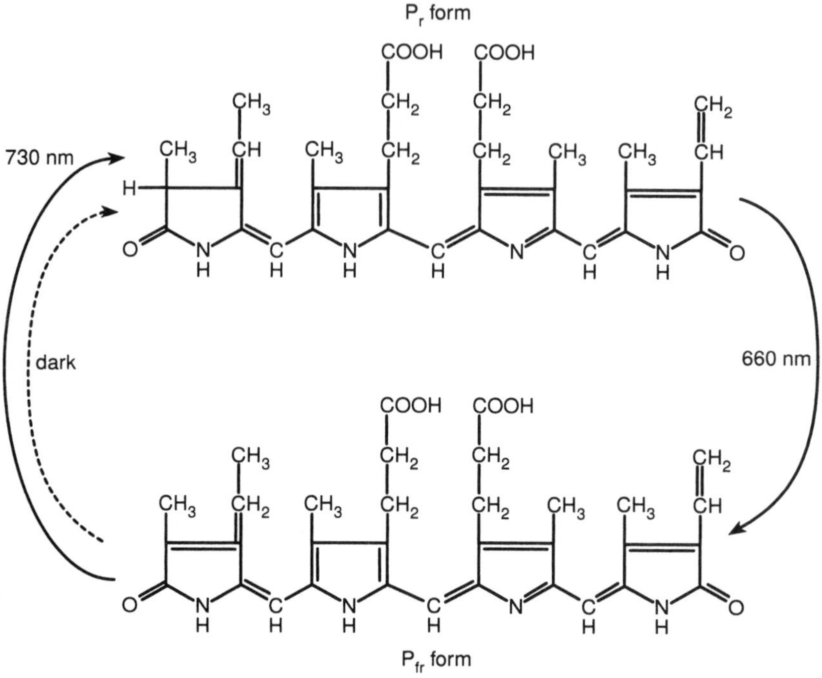

P$_r$ form

730 nm

dark

660 nm

P$_{fr}$ form

Fig. 1.17 The light-triggered interconversion of the 'red (P_r)' and 'far-red (P_{fr})' forms of the phytochrome chromophore. In the dark, the P_{fr} form slowly reverts to the P_r form. This is the biological switch mechanism which controls phototropism, release of seed dormancy, induction of flowering and a number of light-sensitive physiological processes, all at low light intensities often insufficient for photosynthesis.

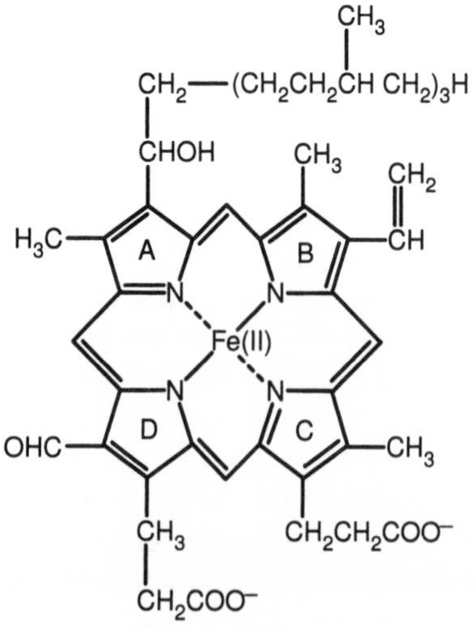

(18)

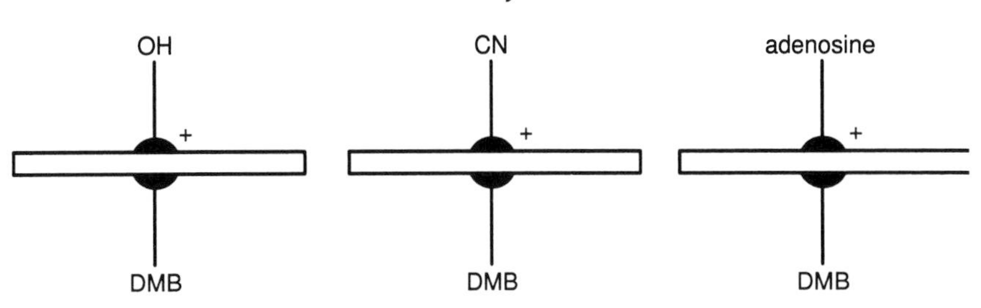

Fig. 1.18 Diagrammatic representation of the structure of the common forms of vitamin B_{12}: (a) hydroxycobalamine, the form mostly found in the diet; (b) the cyanocobalamine form, as isolated for pharmaceutical purposes; and (c) the functional, i.e. coenzymic form of vitamin B_{12}. The flat planar structure represents that of the corrin ring with the cobalt ion, shown in black, at its centre. DMB is the 5,6-dimethylbenzimidazole unit. For details see structure 11.

Many patients, though not all, responded to a treatment which initially consisted of eating large amounts of raw liver, and later, to taking liver concentrates. Castle showed that the reason why some patients did not respond was that in addition to the dietary intake of the then unknown food factor, a second essential factor was involved. This was found to be a special carrier glycoprotein, designated by Castle as 'intrinsic factor' to distinguish it from the dietary 'extrinsic factor'. Produced by the gastric mucosa, the carrier glycoprotein is needed to complex the dietary factor, now known to be vitamin B_{12} , and so facilitate its transport through the stomach wall and into the blood stream. Once there, it rapidly releases its vitamin B_{12} for use by various tissues. Today, patients presenting with pernicious anaemia are, almost without exception, individuals unable to produce the required intrinsic factor and in consequence they have to be treated by intramuscular injection of vitamin B_{12}.

As the clinical term 'pernicious anaemia' indicates, insufficiency of the vitamin prevents normal formation of erythrocytes and patients display the characteristic pallor and weakness of all anaemias. If not treated, the patient develops 'achlorhydria', an inability to produce gastric HCl and this results in digestive problems and vomiting. In chronic, untreated cases, symptoms often include a persecution complex, and a degeneration of the central nervous system which progressively becomes irreversible.

In its coenzymic form, vitamin B_{12} functions as a methyl carrier in the enzyme-catalysed methylation of homocysteine to form methionine (Fig. 1.19). This methyl group originates from 5-methyltetrahydrofolate (Section 7.5). A second coenzymic role of vitamin B_{12} is the isomerization of methylmalonyl CoA to succinyl CoA. Methylmalonyl CoA arises as a catabolic product of the pyrimidine thymine and of

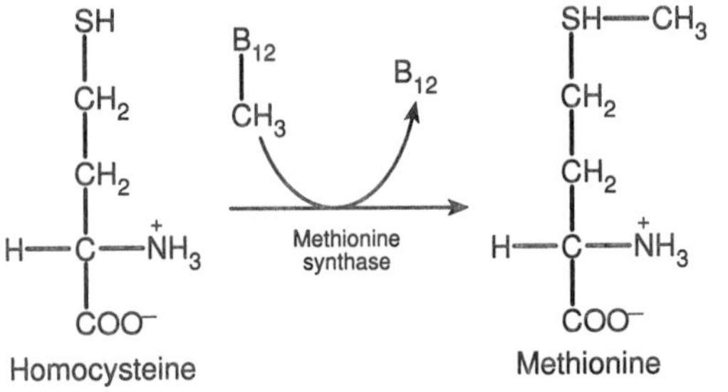

Fig. 1.19 Coenzymic vitamin B_{12} functions as a methyl carrier in the enzyme-catalysed methylation of homocysteine to produce methionine.

the amino acids methionine, isoleucine and valine. Its isomerization (Fig. 1.20) is catalysed by methylmalonyl CoA mutase, which requires vitamin B_{12} as its coenzyme. Succinyl CoA, the product of the enzymic reaction, is further catabolized, oxidatively, by the tricarboxylic acid (Krebs) cycle. Vitamin B_{12} deficiency results in impairment of the isomerization of methylmalonyl CoA to succinyl CoA and causes methylmalonate to appear in significant, detectable, amounts in the urine; another of the characteristic features of pernicious anaemia.

1.9 BIOLOGICAL AND PHARMACOLOGICAL ACTIVITY

Although of substantial biochemical importance, the naturally occurring pyrroles and porphyrins have little direct biological activity. Two weakly antibacterial pyrrole antibiotics have been described, pyrrolni-

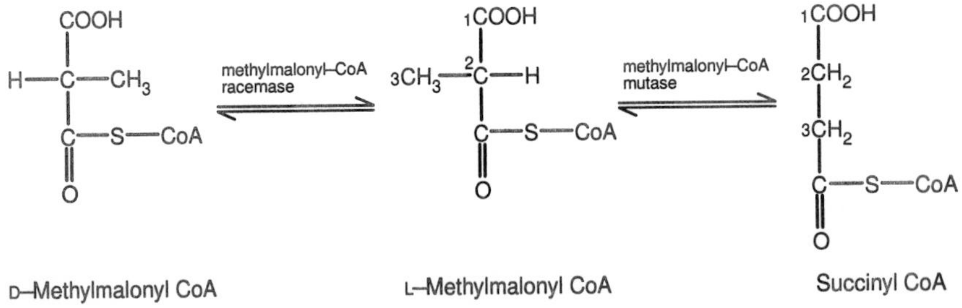

Fig. 1.20 D-Methylmalonyl CoA, arising from the catabolism of methionine, isoleucine, valine and thymine, is racemized by methylmalonyl-CoA racemase to the L-isomer. This is then isomerized, by the coenzyme B_{12}-dependent enzyme methylmalonyl-CoA mutase, to succinyl-CoA. Patients suffering from vitamin B_{12} deficiency (pernicious anaemia) consequently excrete significant, detectable amounts of methylmalonic acid in the urine; this is a diagnostic feature of the disease.

trin (19) and pyoluteorin (20). The former is of more clinical interest since it exhibits activity against some pathogenic species of fungi, including those of *Epidermophyton, Microspora,* and *Trichophyton*. Another naturally occurring pyrrole, the methyl ester of 2-carboxy-4-methylpyrrole (21) has been identified as an insect pheromone. Reduced pyrroles are represented in this category by the tobacco alkaloid nicotine (22) which contains a pyrrolidine ring. Nicotine is a vasoconstrictor, acting by stimulating the release of vasopressin and adrenalin. Two other well-known examples of pyrrolidine-containing alkaloids with pharmacological activity are cocaine and hyoscyamine.

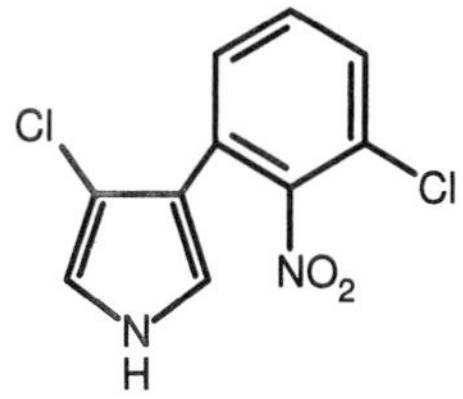

(19)

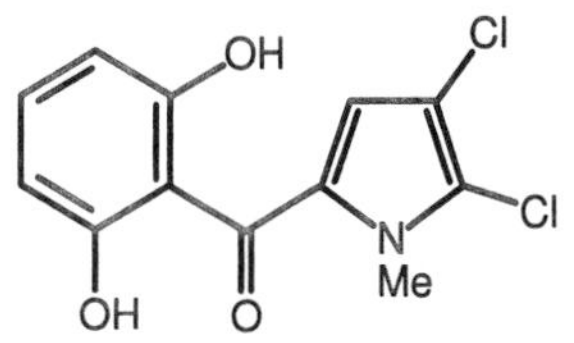

(20)

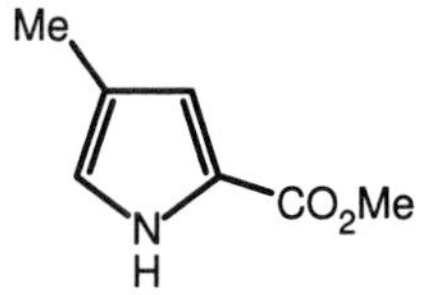

(21)

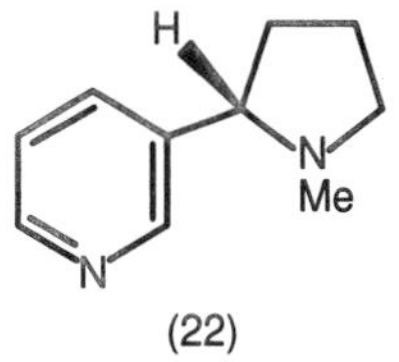

(22)

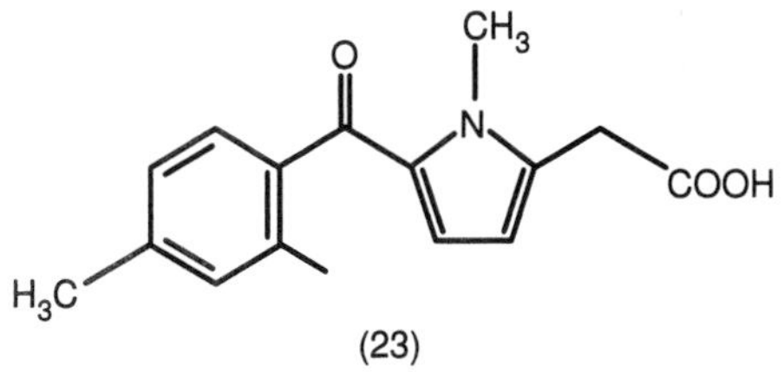

(23)

1.10 SYNTHETIC COMPOUNDS OF CLINICAL IMPORTANCE

As might be expected from the relative absence of biological activity amongst the pyrroles, there are few synthetic pyrroles of clinical importance. One exception is Tolmetin (23) which is currently used in treating rheumatoid arthritis. In this connexion, it has received clinical comparison with aspirin and with indomethacin but does not have the same tendency to cause lesions of the digestive tract. Like aspirin, Tolmetin appears to function as an inhibitor of prostaglandin synthase and hence is an anti-inflammatory agent.

REFERENCES

Battersby, A.R., Fookes, C.J.R., Matcham, G.W.J. and McDonald, E. (1980) Biosynthesis of the pigments of life: formation of the macrocycle. *Nature* **285**, 17–21.

Battersby, A.R. (1985) Biosynthesis of the pigments of life. *Proc. Roy. Soc. London B*, **225**, 1–26.

Battle, A.M. del C. and Stella, A.M. (1978) *International Journal of Biochemistry* **9**, 861–864.

Beale, S.E., Gough, S.P. and Granick, S. (1975) Biosynthesis of δ-aminolevulinic acid from the intact carbon skeleton of glutamic acid in greening barley. *Proc. Natl. Acad. Sci. USA*, **72**, 2719–2723.

Bogorad, L. (1958) The enzymatic synthesis of porphyrins from porphobilinogen. *J. Biol. Chem.*, **233**, 510–515.

Cookson, G.H. and Rimington, C. (1954) Porphobilinogen. *Biochemical Journal*, **57**, 476–484.

Falk. J.E., Dresel, E.I.B. and Rimington, C. (1953) Porphobilinogen as a porphyrin precursor and interconversion of porphyrins in a tissue system. *Nature*, **172**, 292–294.

Jackson, A.H., Sancovich, H.A., Ferramola, A.M., Evans, N., Games, D.E., Matlin, S.H., Elder, G.H., Smith, S.G. (1976) Macrocyclic intermediates in the biosynthesis of porphyrins. *Philosoph. Trans. Roy. Soc. London* (Biol) B, **273**, 191–206.

Lemberg, R. and Legge, J.W. (1949) in *Haematin Compounds and Bile Pigments*, Wiley-Interscience, New York.

Shemin, D. and Rittenberg, C. (1945) Utilization of glycine for the synthesis of a porphyrin. *J. Biol. Chem.* **159**, 567–568.

Westall, R.G. (1952) Isolation of porphobilinogen from the urine of a patient with acute porphyria. *Nature*, **170**, 614–616.

WIDER READING (BOOKS AND REVIEWS)

Abeles, R. and Dolphin D. (1976) The Vitamin B_{12} Coenzyme. *Accounts Chem. Res.*, **9**, 114–120.

Amesz, J. (ed) (1987) *Photosynthesis*, Elsevier, Amsterdam.

Chadwick, D.J. and Ackrill, K. (eds) (1944) The Biosynthesis of the Tetrapyrole Pigments. CIBA symposium. J. Wiley, Chichester.

Dolphin D. (ed) (1982) *B_{12}*, Vols 1 and 2, Wiley-Interscience, New York.

Dolphin, D. (ed) (1978) *The Porphyrins*, Vols 1–7, Academic Press, New York.

Glazer, A.N. (1983) Comparative biochemistry of photosynthetic light-harvesting systems. *Annual Review of Biochemistry*, **52**, 125–157.

Halpern, J. (1985) Mechanisms of coenzyme B_{12}-dependent rearrangements. *Science*, **227**, 869–875.

Jahn, D., Verkamp, E. and Söll, D. (1992) Glutamyl transfer RNA: A precursor of haem and chlorophyll biosynthesis. *Trends in Biochemical Sciences*, **17**, 215–218.

Rimington, C. (1989) Haem biosynthesis and porphyrins: 50 years in retrospect. *J. Clin. Chem. Clin. Biochem.*, **27**, 473–486.

Scheer, H. (ed) (1991) *Chlorophylls*, CRC Press, Boca Raton, FL.

Vierstra, R.D. and Quail, P.H.(1986) Phytochrome the protein, in *Photomorpho-*

genesis of Plants (eds R.E. Kendrick and G.H.M. Kronenberg), Martinus-Nijhoff, Boston, pp.35–60.

Youvan, D.C. and Marrs, B.L. (1987) Molecular mechanisms of photosynthesis. *Scientific American*, **256**, 42–48.

Zuber, H. (1986) Structure of light-harvesting antenna complexes of photosynthetic bacteria, cyanobacteria and red algae. *Trends in Biochemical Sciences*, **11**, 414–419.

CHAPTER 2

Imidazoles and benzimidazoles

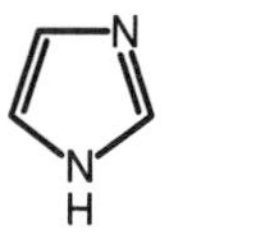

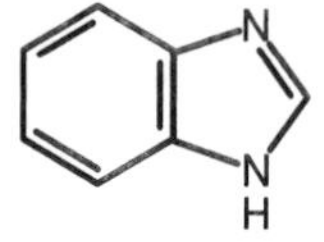

2.1 DISCOVERY AND NATURAL OCCURRENCE

The parent compound imidazole (1) was first synthesized in 1858 by Debus but imidazole derivatives were known before this, e.g. lophine (2,4,5-triphenylimidazole) (2) was synthesized in 1845 by Laurent.

(1)

(2)

The imidazole nucleus appears in a number of naturally occurring compounds. The amino acid histidine (3), for example, is an important constituent of proteins and hence of many enzymes, in which it often forms part of the catalytic site. For example, a number of proteases, such as chymotrypsin, trypsin and elastase, each have a histidine residue that functions in their respective catalytic processes (Section 2.5). So, too, does ribonuclease.

(3)

The product of histidine decarboxylation, histamine (4), is a naturally occurring compound with pronounced pharmacological activity (Section 2.6). It is found in normal tissues and blood, and occurs widely in nature as a result of putrefaction, i.e. bacterial degradation of protein and decarboxylation of the liberated histidine. It is a component of the toxin contained in the trichomes (hairs) of the stinging nettle and which is injected under the skin when we come in contact with this plant (*Urtica dioica*). The irritation and blistering caused by histamine is also seen in susceptible people following trauma, stings or insect bites and is known as 'nettle-rash' or, more formally, 'urticaria'. Generalized urticaria is often due to sensitivity to foods such as strawberries and shellfish, or to drugs such as penicillin.

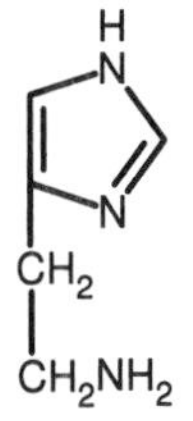

(4)

Pilocarpine (5), the betaine hercynine (6), and the sulphur-containing derivative ergothioneine (7) are all biologically active imidazole derivatives found in nature. Pilocarpine is the cholinergic principle of *Pilocarpus jaborandi* and related plants, hercynine is present in many fungi, especially in *Amanita muscaria* (fly agaric) and *Agaricus compestris* (edible field mushroom), and has also been found in *Limulus polyphemus* (king crab). Ergothioneine occurs in the ergot fungus *Claviceps pupurea*, and in mammalian blood, semen and other tissues.

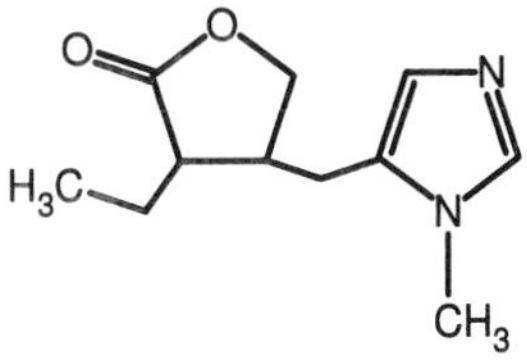

(5)

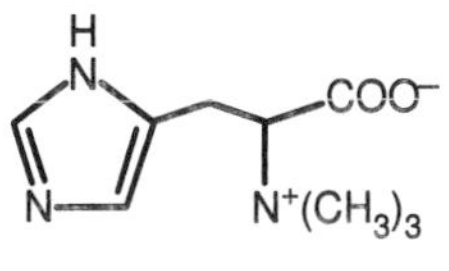

(6)

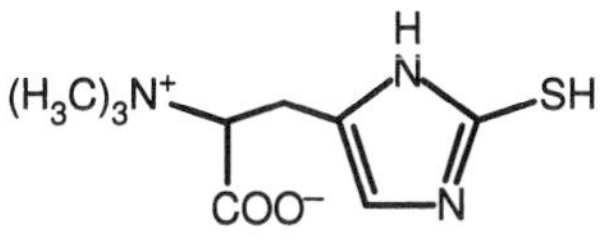

(7)

The ring-fused imidazole benzimidazole (8) was first described in 1878 by Wundt who prepared it by reacting *o*-phenylenediamine with formic acid. It occurs in nature as a structural component of the vitamin B_{12} molecule and several benzimidazoles are commercially available as pharmaceutical, agricultural, and veterinary products.

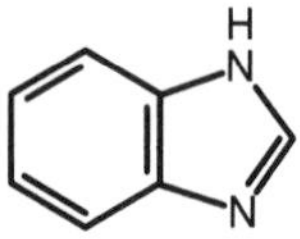

(8)

Vitamin B_{12} (Chapter 1, Section 4.3) exhibits a number of species-related modifications of the benzimidazole component, e.g. the vitamin from mammalian and many microbial sources contains 5,6-dimethyl-1-(α-D-ribofuranosyl)benzimidazole-3-phosphate (9) whereas in the vitamin from some microbial species this is replaced by the α-D-ribofuranosyl-3′-phosphate of 5-methoxybenzimidazole, or of 5-methylbenzimidazole. In some species it is the ribotide of benzimidazole itself that is involved.

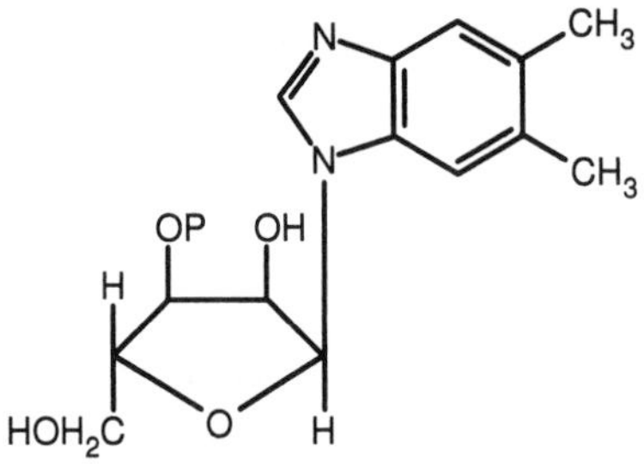

(9)

Biotin, another member of the B-group of water-soluble vitamins, also contains a fused imidazole ring within its molecule (10). In this case, the other ring is that of thiophene. In the context of fused imidazole ring

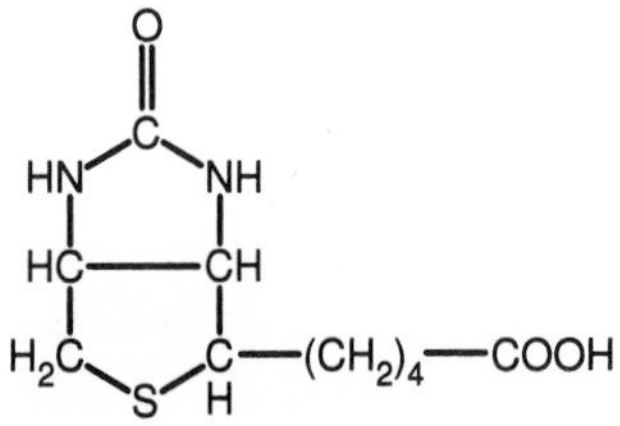

(10)

compounds, it should also be remembered that purines (Chapter 6) are imidazopyrimidines.

2.2 CHEMICAL PROPERTIES OF BIOCHEMICAL INTEREST

Imidazole is a white crystalline solid, readily soluble in water and polar solvents. The molecule is a planar 5-membered ring with three carbon atoms and two nitrogens, the latter 1,3 to one another (1). Contributions of one π-electron from each carbon atom and from the =N- atom, plus two from the –NH– nitrogen, result in an aromatic sextet.

Imidazole is an excellent nucleophile and reacts readily at nitrogen with alkylating and acylating reagents. The *N*-acyl derivatives are good acyl-transfer agents, comparable to acyl halides or acyl anhydrides. Reactions at the carbon atoms of the imidazole ring are, in general, those expected of a stable aromatic heterocycle less susceptible to electrophilic attack than pyrrole; substitution reactions that do not destroy the aromatic character are predominant. Nucleophilic substitution is unlikely unless there is a strongly electron-withdrawing substituent elsewhere in the ring. If this is not the case, carbon-2 is the position most likely to undergo nucleophilic attack. With benzimidazoles, the fused benzene ring provides sufficient electron-withdrawal effect to permit a variety of nucleophilic substitution reactions to take place at carbon-2.

2.3 BIOSYNTHESIS

2.3.1 Histidine and histamine

The biosynthetic pathway leading to production of histidine (Fig. 2.1) involves transfer of *N*-1 and *C*-2 of the adenine moiety of ATP to the ribose phosphate unit of 5-phosphoribosyl-1-pyrophosphate (PRPP). As PRPP is also a key compound in purine and pyrimidine biosynthesis and in the related salvage pathways (Chapters 5 and 6) its involvement here represents a direct link between these processes and histidine biosynthesis.

The transfer of the *N*-1 and *C*-2 of adenine is effected through a condensation reaction between PRPP and ATP, and is followed by a series of reactions involving ring-opening, isomerization, and amino group transfer. During this process, the residue of the ATP molecule is released in the form of 5-amino-4-imidazolecarboxamide ribotide (AICR) which, being an important intermediate in purine biosynthesis, is salvaged.

Studies with bacteria and with yeast indicate that metabolic channelling may be operating in histidine biosynthesis. Evidence in support of

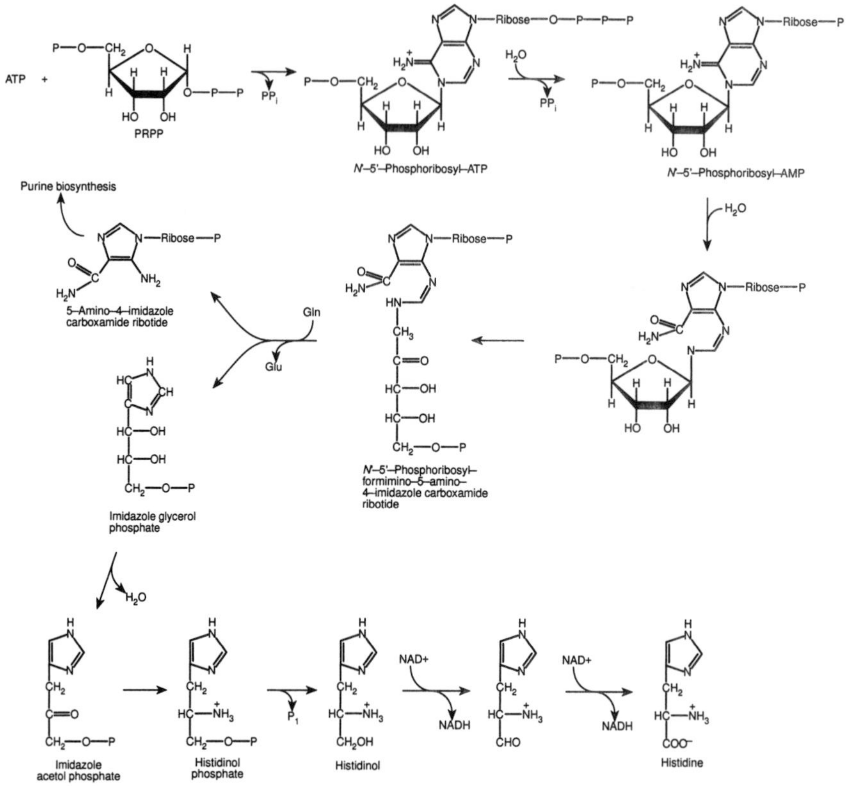

Fig. 2.1 Pathway of histidine biosynthesis from ATP and 5-phosphoribosyl-1-pyrophosphate (PRPP). The first three steps, in effect, result in ring-opening of the pyrimidine ring of adenine. Donation of the amide nitrogen of glutamine is followed by release of 5-amino-4-imidazole carboxamide ribotide (a purine precursor) and the remainder of the molecule is ring-closed to yield imidazole glycerol phosphate. This is converted to histidine by sequential dehydration, transamination, dephosphorylation and dehydrogenation to yield histidine. All the steps are enzyme-catalysed.

this view is that significant pools of intermediates cannot be detected and several of the enzymic activities involved in the biosynthetic scheme shown in Fig. 2.1 appear to be functions of a single protein.

The formation of histamine (Fig. 2.2) from histidine is a typical biochemical decarboxylation, catalysed by a widely distributed pyridoxal 5′ -phosphate-dependent enzyme. As discussed above (Chapter 2) the benzimidazole ring system is found within the molecular structure of vitamin B_{12} and its derivatives (cobalamins). The question therefore arises as to the biological origin of this fused imidazole nucleus. Biosynthetic studies (Renz, 1970) with the bacterium *Propionibacterium shermanii*, which is used in industrial-scale fermentations to produce vitamin

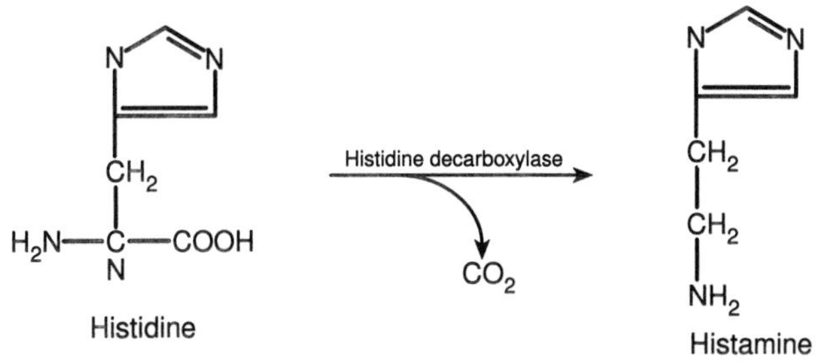

Fig. 2.2 Enzymic decarboxylation of histidine to form histamine.

B_{12}, implicated riboflavin as precursor of the benzimidazole ring system (Fig. 2.3). Later studies with radioisotopic tracers (Alworth *et al.*, 1971) indicated that 6,7-dimethyl-8-ribityllumazine (Fig. 2.3), an intermediate in riboflavin biosynthesis (Section 8.3.1), is the more immediate precur-

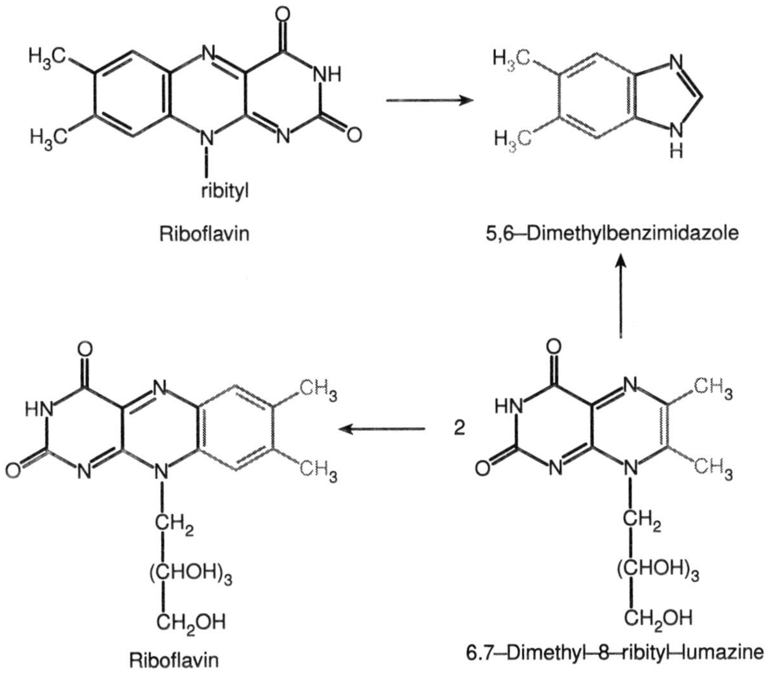

Fig. 2.3 Biosynthetic interrelationship of 5,6-dimethylbenzimidazole, 6,7-dimethyl-8-ribitylumazine, and riboflavin. Biosynthetic studies with *Propionibacterium shermanii* by Renz (1970) implicated riboflavin as the precursor of 5,6-dimethylbenzimidazole. Later work by Alworth *et al.* (1971) indicated that 6,7-dimethyl-8-ribitylumazine, the known immediate precursor of riboflavin, is also the immediate precursor of 5,6-dimethylbenzimidazole. The evidence for this came from studies with ^{14}C-labelled 6,7-dimethyl-8-ribitylumazine. The traced biosynthetic fate of the ^{14}C-atoms from the precursor is shown by the shaded structures.

sor and that the 4,5-dimethyl-1,2-phenylene structural unit of 5,6-dimethylbenzaldehyde is derived from the same type of biomolecular condensation that produces ring A of riboflavin (Fig. 8.4). This involves two molecules of 6,7-dimethyl-8-ribityllumazine, each of them donating a four-carbon unit to the assembly of the 8-carbon dimethylphenylene structure present in both riboflavin and 5,6-dimethylbenzimidazole.

2.4 CATABOLISM

The catabolism of histidine (Fig. 2.4) begins with the attack of a specific lyase which splits out ammonia to yield urocanic acid. This is followed by a simultaneous reduction and internal oxidation involving the elements of water and catalysed by the enzyme urocanase. The product, 4-imidazolone-5-propanoate, is ring-opened to yield formiminoglutamate which, in turn, is hydrolytically cleaved by a tetrahydrofolate-dependent enzyme. During the latter process, the formimino group is transferred as a one-carbon unit to the coenzymic acceptor tetrahydrofolic acid (Chapter 7).

2.5 BIOLOGICAL FUNCTIONS

In the bioactive proteins of which the imidazole amino acid histidine is a component, histidine residues often play a key part in the functioning of the protein concerned. For example, studies of the mechanism of oxygen binding by haem proteins, such as haemoglobin and myoglobin, implicate histidine residues in the biochemical mechanism. The prosthetic haem group of haemoglobin is non-covalently bonded into a hydrophobic groove in the haemoglobin molecule. Within the haem unit, Fe(II) is chelated at the centre of the planar protoporphyrin IX component where it is held by 2 covalent and 2 coordinate bonds, involving the four *N*-atoms of the porphyrin. Fe(II) has a coordination number of six and it has four ligands within the protoporphyrin IX molecule (the four pyrrole nitrogen atoms of the porphyrin macrocycle). The remaining two coordination sites lie along an axis perpendicular to the plane of the ring (Fig. 2.5). One of these is occupied by a globin histidine residue, known as the 'proximal histidine'. The remaining site is the oxygen-binding site, located on the other side of the planar porphyrin ring, with a second histidine residue (the 'distal histidine') near to the haem but not bonded to it. The O_2 molecule to be transported is attached to the iron atom and sandwiched between it and the ring nitrogen of the distal histidine (Fig. 2.5). The special feature of this system, biochemically, is that it permits the binding of oxygen to the iron atom but prevents the usual consequence of such an event, i.e. oxidation of the metal.

Histidine

Histidase → $\overset{+}{N}H_4$

Urocanate

Urocanase, H_2O

4–Imidazolone–3–propionate

Hydrolase, H_2O

N–Formimino–L–glutamate

Glutamate formimino–transferase

Tetrahydrofolate → N^5–formiminotetrahydrofolate

Glutamate

Fig. 2.4 Catabolism of histidine. In a series of enzyme-catalysed reactions, the ureido carbon atom of the imidazole ring is removed as a one-carbon unit, leaving glutamate as the product.

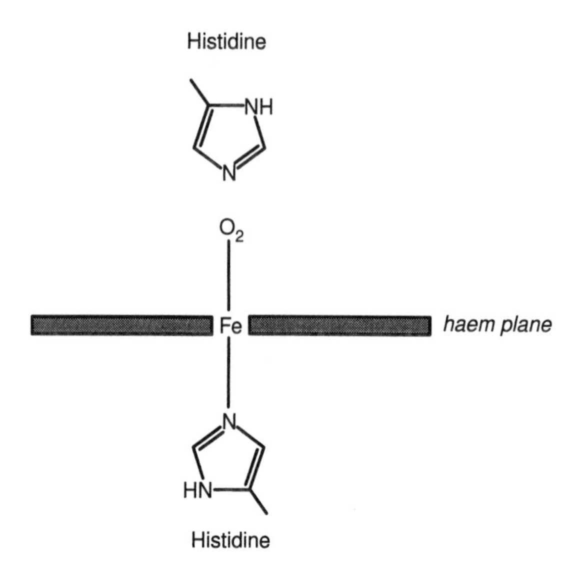

Fig. 2.5 Model of the oxygen-binding site of haemoglobin. The upper histidine residue in the diagram is the distal histidine (E7) of the globin moiety; the lower one is the F8 histidine residue of globin. At the centre of the planar protoporphyrin IX unit of haem, the iron atom is in the Fe(II) state and is the point of attachment of the O_2 molecule carried by haemoglobin.

Pancreatic ribonuclease A, which catalyses the hydrolysis of ribonucleic acid, also depends for its activity on histidine residues at its catalytic site. Treatment of ribonuclease with the alkylating agent iodoacetate results in carboxymethylation of the imidazole ring of histidine 119 or histidine 12 but not both in the same molecule. Modification of either histidine residue inactivates the enzyme. In the presence, however, of substrate, or of competitive inhibitors which can be regarded as substrate analogues, the enzyme is protected from the iodoacetate effect. Studies of the pH optimum of ribonuclease had earlier led to the suggestion that two histidine residues participate in the catalysis, one in the basic form and the other in the acidic form. Taken together, these observations indicate that the two histidine residues 12 and 119 are in proximity to the catalytic site and act in concert as proton donor and proton acceptor in the catalytic hydrolysis of RNA shown in Fig. 2.6.

The digestive enzyme carboxypeptidase A also possesses functional histidine residues. This enzyme hydrolyses the carboxy-terminal peptide bond in polypeptides, especially when there is an aromatic or bulky aliphatic side chain attached to the terminal amino acid residue. A zinc ion, essential for the catalytic activity, is located in a crevice near the surface of the protein and it is coordinated to two histidine residues

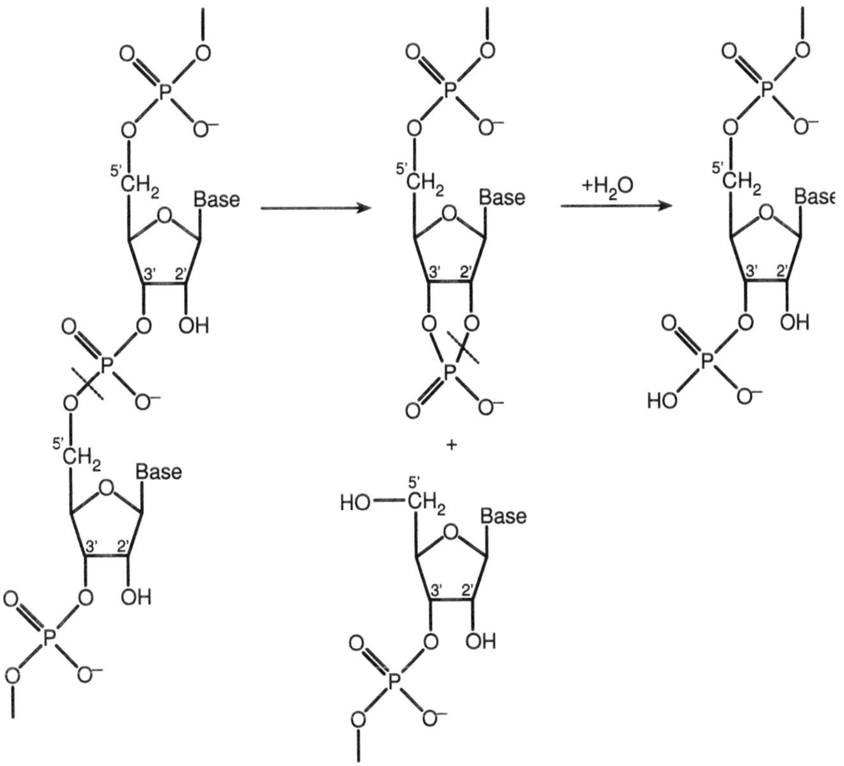

Fig. 2.6 Hydrolysis of RNA catalysed by pancreatic ribonuclease (ribonuclease A). Experimental evidence indicates that two histidine residues (12 and 119) in the enzyme molecule are in close proximity to the catalytic site and act in concert as proton donor and proton acceptor during the hydrolysis.

(histidines 69 and 196) a glutamate residue, and a water molecule. Alcohol dehydrogenase, too, has a zinc ion attached to a histidine residue (histidine 67) at its catalytic site.

Chymotrypsin is another protease with a histidine residue (histidine 57) involved in its catalytic site. During catalysis of the hydrolysis of a peptide bond by this enzyme (Fig. 2.7), the imidazole ring of the histidine residue accepts a proton from the adjacent serine 195 residue. The protonated ring is stabilized by electrostatic interaction with the $-COO^-$ group of the nearby aspartate 102 residue. This tetrahedral transitional state (Fig. 2.7) also has the effect of precisely positioning the imidazole ring of histidine 57 so that the proton held by the protonated ring is donated to the nitrogen atom of the peptide bond to be cleaved. The amino component of the peptide bond then diffuses away leaving an acyl-enzyme intermediate.

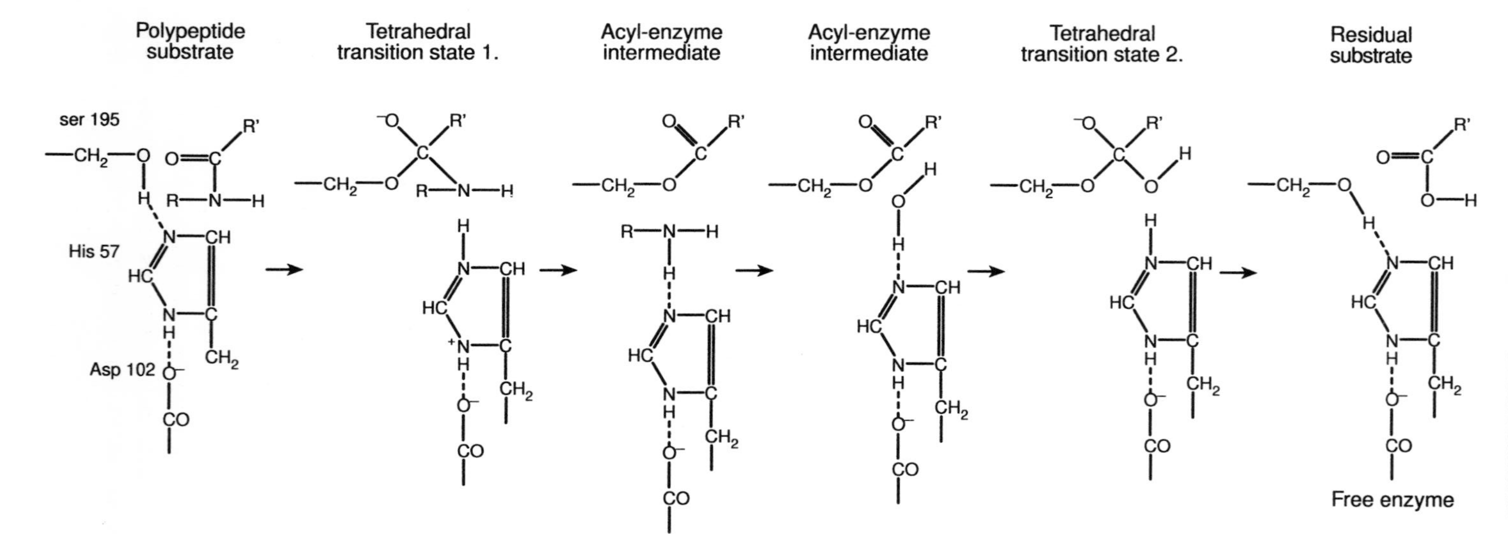

Fig. 2.7 Hydrolysis of a peptide by chymotrypsin, showing the involvement of the histidine 57 residue of the enzyme. The initial acylation stage involves formation of a tetrahedral transition state (1) and cleavage of the peptide bond. The amine product diffuses away and is replaced by a molecule of water. In effect, the process reverses *via* a second tetrahedral transition state (2) and deacylation occurs, releasing the enzyme and the residue of the hydrolysed polypeptide substrate.

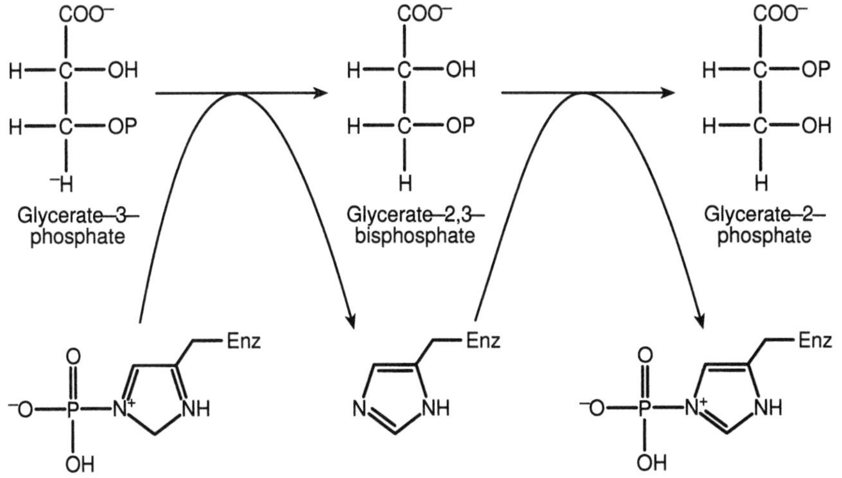

Fig. 2.8 Role of the histidine residue at the catalytic site of the enzyme phosphoglycerate mutase. By acting as donor and acceptor of a phosphoryl group, the imidazole ring catalyses the isomerization of glycerate 3-phosphate to glycerate 2-phosphate.

Deacylation of the enzyme (Fig. 2.7) begins when a water molecule replaces the lost amino component of the hydrolysed substrate. A proton is withdrawn from the water molecule by histidine 57 and the OH^- ion, so formed, attacks the carbonyl carbon of the acyl group attached to serine 195, resulting in a transitional state. Finally, histidine 57 donates the proton to the oxygen atom of serine 195 and this releases the substrate residue from the enzyme (Fig. 2.7).

In the course of glycolysis, the enzyme phosphoglycerate mutase catalyses the isomerization of 3-phosphoglycerate to 2-phosphoglycerate. This enzyme, isolated from yeast and from rabbit muscle, has been shown to form phosphoenzyme intermediates in which the imidazole ring of a histidine residue features prominently. In the prevalent physiological form of phosphoglyceromutase, a phosphoryl group is covalently bound to *N*-3 of the imidazole ring of a histidine residue at the catalytic site (Fig. 2.8). The enzyme requires 2,3-bisphosphoglycerate as a cofactor and under physiological conditions the reaction catalysed (Fig. 2.8) involves only a small free energy change ($\Delta G^{o\prime}$ = 0.83 kJ/mol).

A further example of a biochemical reaction dependent on the catalytic involvement of the imidazole ring of a histidine residue is the conversion of succinyl-CoA to succinate, seen in the tricarboxylic acid cycle. Hydrolysis of succinyl-CoA to succinate is facile; succinyl-CoA has a high free energy of hydrolysis ($\Delta G^{o\prime}$-8kcal/mol). The energy which would otherwise be dissipated by this process is, however, con-

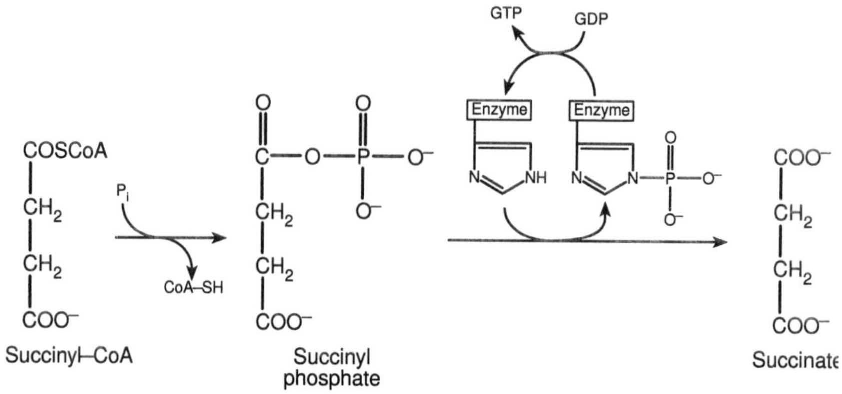

Fig. 2.9 The energy-conserving succinyl-CoA synthetase reaction in the tricarboxylic acid cycle. Succinyl phosphate, formed from succinyl-CoA, phosphorylates *N*-3 of the imidazole ring of a histidine residue in the enzyme protein. In turn, this phosphorylates GDP to GTP, the guanosine analogue of ATP, and the enzyme is restored to its original state. Succinyl-CoA is thus converted to succinate and the high $\Delta G^{o\prime}$ of hydrolysis is conserved by the generation of the high energy compound GTP.

served by the concomitant phosphorylation of GDP to GTP, catalysed by succinyl-CoA synthetase. This enzyme functions through the reversible phosphorylation of a histidine residue in the protein structure (Fig. 2.9).

Like most other water-soluble vitamins of the B-group, biotin functions as a coenzyme. Structurally, its molecule consists of a fused imidazole and thiophene ring. The imidazole moiety functions as a carrier of $-CO_2^-$ during enzymic carboxylation reactions (Fig. 2.10) whereas the thiophene moiety anchors the coenzyme, via its carboxylic acid side-chain, to a lysine residue at the catalytic site of the enzyme. Fig. 2.10 shows the biotin-dependent carboxylation of pyruvate to oxaloacetate, catalysed by pyruvate carboxylase. The enzyme-bound biotin is first carboxylated at *N*-1 of its imidazole ring in an ATP-requiring step catalysed by biotin carboxylase. In the second reaction (Fig. 2.10) the carboxyl group is transferred to the acceptor, pyruvate in the example given. A similar mechanism is involved in carboxylating acetyl-CoA to form malonyl-CoA during fatty acid synthesis.

In nature, the fused imidazole benzimidazole is not often seen; nevertheless, 5,6-dimethylbenzimidazole is an important component of the structure of vitamin B_{12}. In this molecule *N*-3 of the benzimidazole unit is linked through a coordinate bond with the cobalt atom held at the

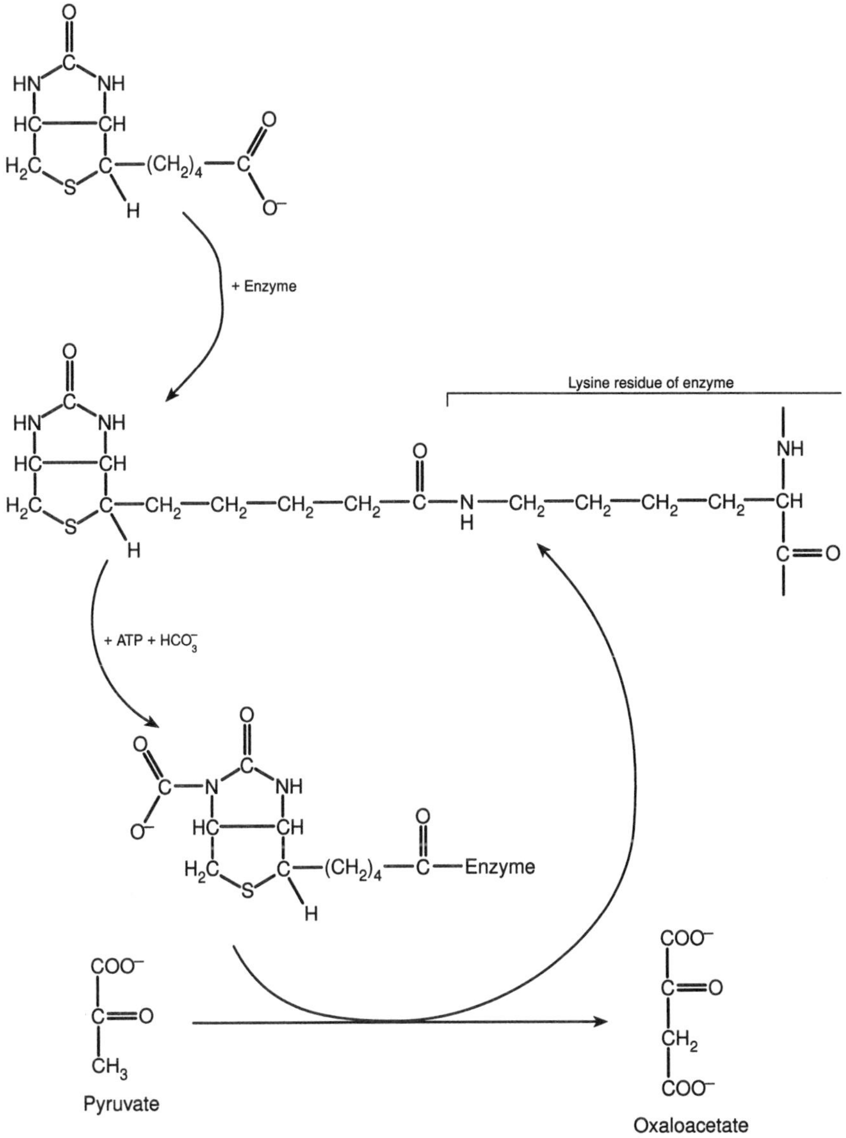

Fig. 2.10 Involvement of the imidazole ring of the coenzyme biotin in enzymic carboxylation. Biotin is bound by the apoenzyme form of pyruvate carboxylase and the *N*-1 atom of the imidazole ring is carboxylated by HCO_3^- in the presence of ATP. After transporting the carboxy group to a molecule of the acceptor pyruvate, to form oxaloacetate, the holoenzyme is ready to repeat the process.

centre of the planar corrin ring (Chapter 2, Structure 11). Three types of reactions are catalysed by a coenzymic form of vitamin B_{12}: (a) methylations, as seen in the synthesis of methionine, (b) intramolecular rearrangements, such as the isomerization of malonyl-CoA to succinyl-CoA, and (c) reduction of the ribose component of a ribonucleotide to yield a 2′-deoxyribonucleotide. During these processes, the cobalt to carbon bond undergoes homolytic fission, leaving the metal bonded to the four nitrogen atoms of the corrin ring and to *N*-3 of the benzimidazole unit. Recently observed conformational changes in the crystalline structure when the enzyme binds the coenzyme, suggest that the 5,6-dimethylbenzimidazole unit moves into a deep cleft in the protein, anchoring the coenzyme tightly. It also appears that during this process, the coordination of the benzimidazole with the cobalt atom may be replaced by coordination of the cobalt with the nitrogen atom of an imidazole ring within a histidine residue in the enzyme protein. The exact function of the benzimidazole unit, however, remains obscure at this time.

2.6 BIOLOGICAL AND PHARMACOLOGICAL ACTIVITY

Histamine (4) is a potent vasodilator. It stimulates secretion of the gastric juice, i.e. the digestive enzymes and HCl, and causes contractions of the smooth muscles of the digestive tract, the uterus, and the bronchi. Local application causes reddening of the skin and nettle-rash, and it is a constituent of bee venom, and of nettle hairs. Few other naturally occurring imidazoles, however, have obvious biological activity.

Synthetic imidazoles, on the other hand, constitute a large group of drugs and medicines, and many of the fungicides currently in clinical use are imidazoles. Examples (Fig. 2.11) include Clotrimazole, Econazole, Isoconazole, Ketoconazole, Miconazole, and Sulconazole. Similarly, a significant number of antiprotozoal agents listed in current pharmacopoeias are imidazole derivatives (Fig. 2.12). These include Benznidazole, Metroniadzole, Tinidazole, and Nimorazole. Another group of synthetic imidazoles are widely used medicinally as antihypertensives, examples of which are shown in Fig. 2.13.

The benzimidazoles are an interesting group of heterocycles in terms of biological and pharmacological activity. As discussed in Sections 2.1 and 2.3 of this Chapter, 5,6-dimethylbenzimidazole is a structural component of the vitamin B_{12} molecule and arises biosynthetically from a side branch of the pathway of riboflavin biosynthesis. It has a marked inhibitory effect on *in vitro* haem biosynthesis but whether this activity has any biological significance is not known. Few, if any, other benzimidazoles occur in nature but synthetic benzimidazoles have a variety of biological and pharmacological effects.

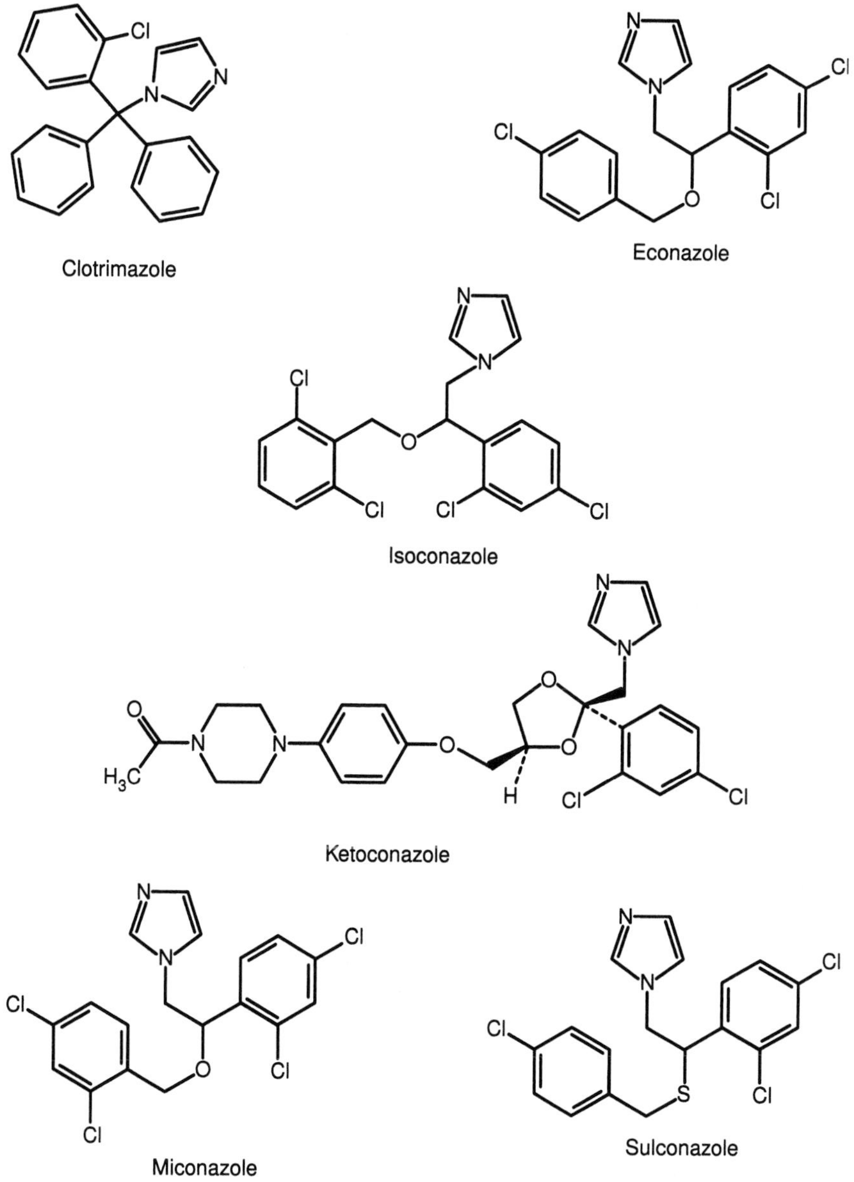

Clotrimazole

Econazole

Isoconazole

Ketoconazole

Miconazole

Sulconazole

Fig. 2.11 Examples of antifungal synthetic imidazoles in current clinical use.

Wooley (1944) showed that benzimidazole suppresses the growth of yeast but that the effect could be reversed by adenine or guanine. He concluded that benzimidazole acts as a purine analogue inhibitor. However, later observations by others indicate that the mechanisms are more complex than this would suggest. Benzimidazole blocks the

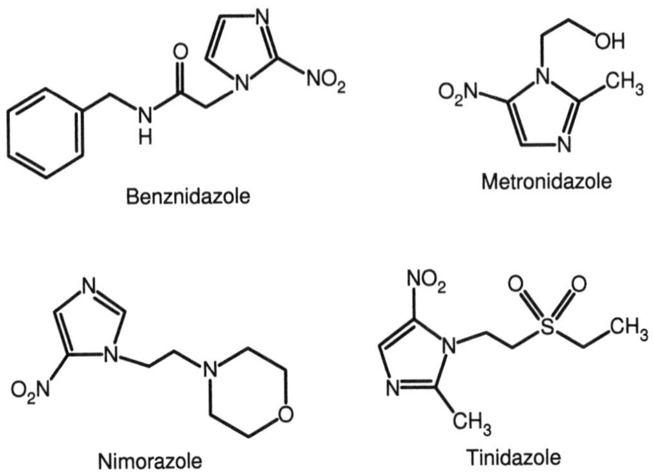

Benznidazole

Metronidazole

Nimorazole

Tinidazole

Fig. 2.12 Some antiprotozoal synthetic imidazoles in current clinical use.

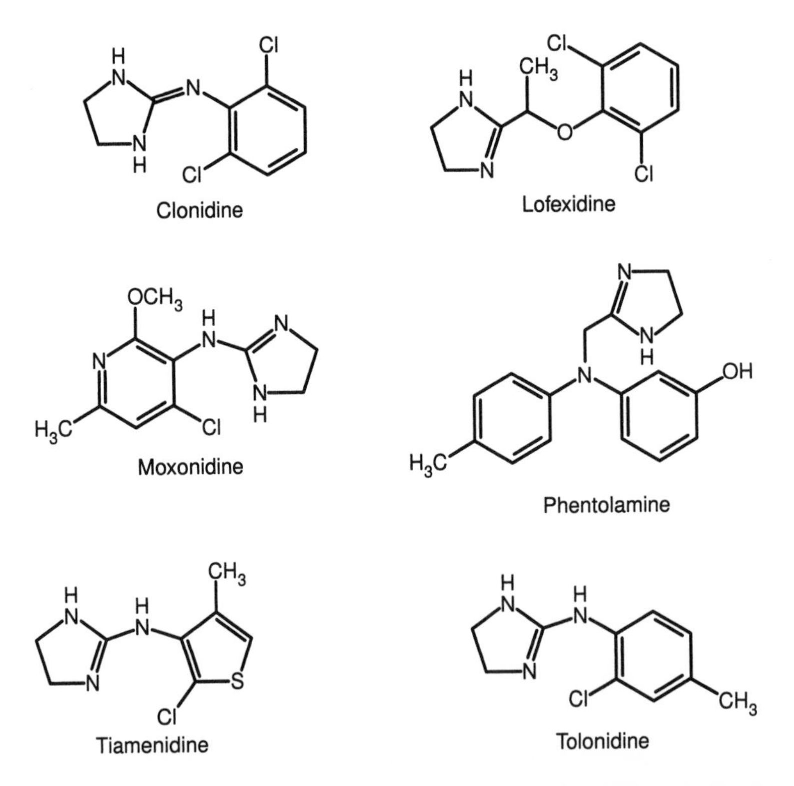

Fig. 2.13 Synthetic imidazoles currently in clinical use as antihypertensives.

oxygen-induced synthesis of respiratory chain enzymes by yeast during the adaptation of anaerobically grown cells to aerobic conditions, and also inhibits protein synthesis. Also with yeast cells, benzimidazole inhibits the derepression of a proline transport system induced by nitrogen-starvation.

Brown & Evans (1979) found that yeast cells inhibited by benzimidazole accumulate hypoxanthine with an associated effect on xanthine, and that the cells became depleted of CMP and UMP. Yeast cultures were shown to ribotylate benzimidazole, and studies using ^{14}C-labelled purines and amino acids led to the conclusion that benzimidazole inhibits yeast growth by competing for 5-phosphoribosyl-1-pyrophosphate (PRPP), so depriving other ribotylation processes such as *de novo* purine and pyrimidine biosynthesis, and the 'salvage pathways' (Chapters 5 and 6).

In higher plants, benzimidazole often exhibits cytokinin activity, such as delaying the senescence of detached leaves. The compound also possesses antiviral properties, and is a bacteriostat. Effects are seen, too, with animal tissues, e.g. the development of chick and frog embryo is suppressed by benzimidazole.

The synthetic benzimidazoles Astemizole (11) and Clemizole (12) are currently used in medicine as antihistamines and antiallergy agents.

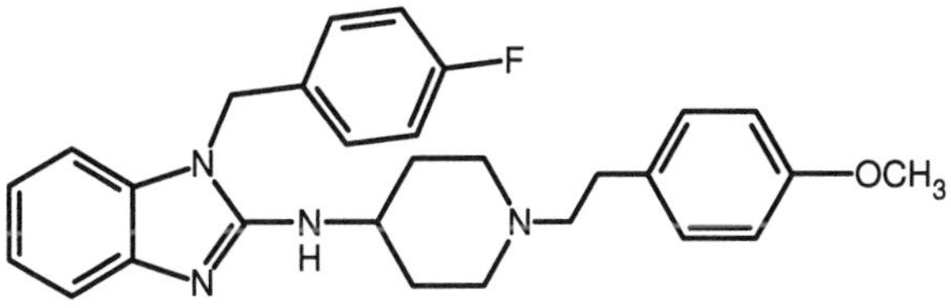

(11)

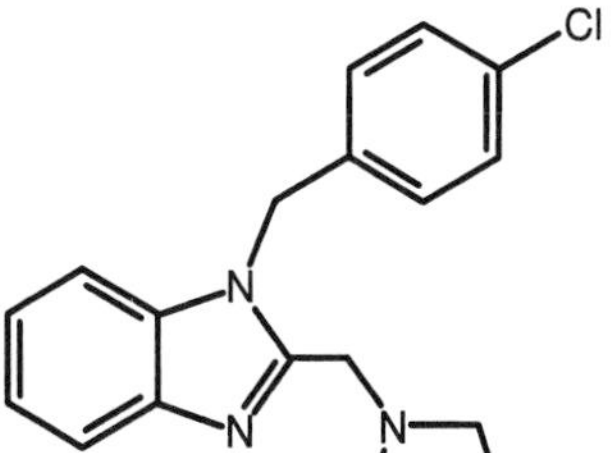

(12)

One of the most successful agricultural and horticultural fungicides of the last 30 years has been methyl 1-(butylcarbamoyl)-2-benzimidazolecarbamate (13), marketed by Du Pont as 'Benomyl' and by ICI as Benlate. This benzimidazole has high activity against a great many fungi including the powdery mildews which affect almost all species of crop plants. Of particular commercial importance are the mildew infections of vine, apple, cucumber and cereals, and Benomyl has been a major advance in treating these crops. Studies of the mode of action of Benomyl have led to the conclusion that the active compound is not Benomyl *per se* but its degradation product methyl-2-benzimidazolecarbamate (14) also known as Carbendazim.

(13) (14)

A second broad-spectrum, systemic benzimidazole fungicide is 2-(4-thiazolyl)-1*H*-benzimidazole (Thiabendazole; 15). This is used for spoilage control of citrus fruit and for treatment and control of Dutch Elm disease in trees. It is also used for the control of fungal infections in seed potatoes. Originally the compound was introduced as an antihelminthic and is still used for this purpose in veterinary medicine.

(15)

REFERENCES

Alworth, W.L., Lu, S-H. and Winkler, M.F. (1971) *Biochemistry*, **10**, 1421–1424.

Bryan, J.K. (1990) Advances in the biochemistry of amino acid biosynthesis, in *The Biochemistry of Plants*, Vol. 16 (eds B.J. Miflin and P.J. Lea), Academic Press, San Diego, pp. 161–195.

Renz, P. (1970) Riboflavin as precursor in the biosynthesis of the 5,6-dimethylbenzimidazole moiety of vitamin B_{12}. *FEBS Letters*, **6**, 187–189.

WIDER READING

Baldwin, J. (1980) Structure and cooperativity of haemoglobin. *Trends in Biochemical Sciences*, **5**, 224–228.

Breslow, R. (1991) How do imidazole groups catalyze the cleavage of RNA in enzyme models and enzymes? Evidence from 'negative catalysis'. *Accounts Chem. Res.*, **24**, 317–320.

Dickerson, R.E. and Geis, I. (1983) *Haemoglobin*, Benjamin/Cummings, Menlo Park, Calif.

Findlay, D., Herries, D.G., Mathias, A.P., Rabin, B.R. and Ross, C.A. (1961) The active site and mechanism of pancreatic ribonuclease. *Nature*, **190**, 781.

Page, M.I. (ed) (1987) *Enzyme Mechanisms*, The Royal Society of Chemistry, London.

Sehgal, P.B. and Tamm, I. (1980) Benzimidazoles and their nucleosides. *Antibiotics and Chemotherapy*, **27**, 93–138.

CHAPTER 3

Pyrazoles

3.1 DISCOVERY AND NATURAL OCCURRENCE

Although pyrazole (1) was first prepared and described by Pechmann in 1898, it was almost another 60 years before the natural occurrence of a pyrazole derivative was reported (Shinano and Kaya, 1957; Noe and Fowden, 1959). This was the amino acid β-pyrazol-1-yl-L-alanine (2) an isomer of histidine (3). Together with its γ-L-glutamyl peptide (4) it has been found in the seeds and seedlings of many species of the plant family Cucurbitaceae, to which cucumber, watermelon, squash and pumpkin belong. Evidence for the existence of free pyrazole in biological tissue was obtained, indirectly, by Dunnill and Fowden (1963) who heated a benzene extract of cucumber seeds with serine, pyridoxal and aluminium sulphate, and obtained β-pyrazol-1-ylalanine. It had previously been shown that authentic samples of pyrazole readily undergo this Al^{3+}/pyridoxal-catalysed reaction with serine, and which simulates the action of the enzyme β-pyrazol-1-ylalanine synthase.

(1) (2) (3)

Aside from β-pyrazol-1-ylalanine, there are few reports of the natural occurrence of simple pyrazoles. There are, however, two pyrazole anti-

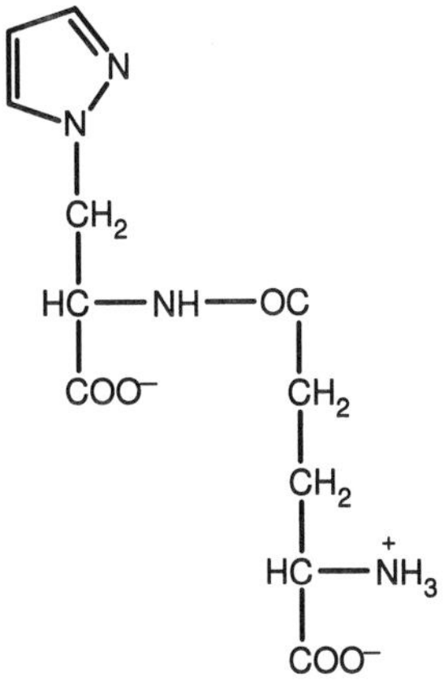

(4)

biotics, pyrazomycin (5) and formycin (6), both produced by the mould *Streptomyces candidus*, and a recent report of the occurrence of 5-oxo-1-*H*-pyrazole in the mushroom *Helvella esculenta*.

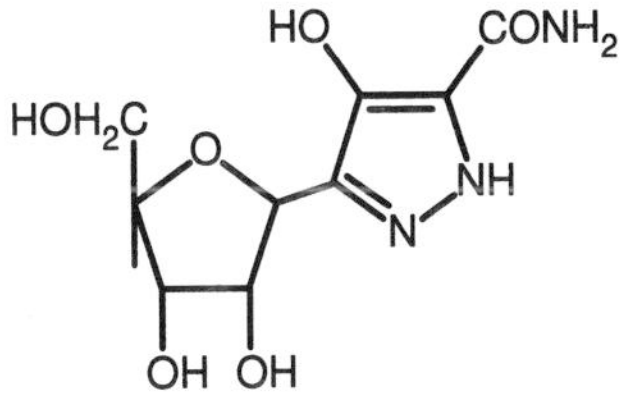

(5)

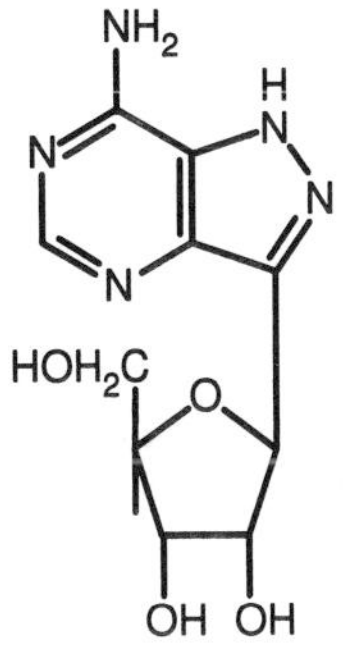

(6)

3.2 CHEMICAL PROPERTIES OF BIOCHEMICAL INTEREST

Pyrazole, itself, is a volatile, colourless, crystalline solid with a pyridine-like odour. It is soluble in water, ethanol, ether and to a lesser extent, benzene and cyclohexane. It is almost insoluble in petroleum ether. As shown in Fig. 3.1, the molecule exhibits tautomerism, and in non-associated solvents it exists primarily as a cyclic hydrogen-bonded dimer. At high concentrations, more than two monomers can associate and evidence has been adduced for the existence of a stable, cyclic trimer.

The three carbon atoms and one nitrogen atom of the pyrazole nucleus contribute four π-electrons. Not being involved in double bond formation, the nitrogen at position 1 also donates its electron pair, so

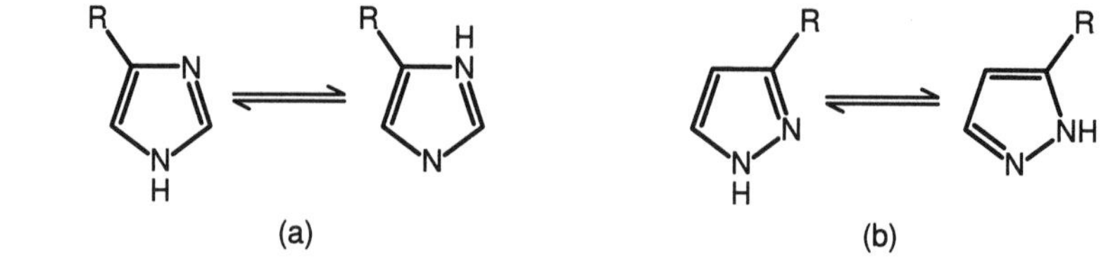

Fig. 3.1 Tautomerism of (a) imidazoles and (b) pyrazoles.

creating an aromatic sextet of π-electrons. Pyrazole is, then, a stable aromatic compound and although amphoteric, more basic than acidic. It is resistant to oxidation and reduction. As a resonance hybrid, electrophilic substitution readily occurs at position 4 in the ring, and alkylation and acylation is possible under moderate conditions.

In the presence of a mild oxidant, pyrazole forms a yellow complex with trisodium pentacyanoaminoferrate. This property has been applied to the quantitative determination of pyrazole in biological extracts (LaRue, 1965).

3.3 BIOSYNTHESIS OF PYRAZOLE AND β-PYRAZOL-1-YLALANINE

The natural occurrence of pyrazole derivatives, albeit small in number, poses an interesting biochemical question. How is a heterocyclic N-N covalent bond formed biosynthetically? The answer has come from recent studies of the biosynthesis of pyrazol-1-ylalanine by tissues of cucumber (*Cucumis sativus*). Incorporation studies with ^{14}C-labelled compounds implicated 1,3-diaminopropane as the precursor of the pyrazole ring (Brown, Flayeh and Gallon, 1982) and this was later confirmed when an enzymic extract of cucumber tissues was shown to catalyse the cyclization of this diamine (Brown and Diffin, 1990). Pyrazole synthase, the enzyme responsible, is FAD-dependent and the reaction catalysed involves a sequential series of dehydrogenations with 2-pyrazoline as an intermediate (Fig. 3.2). The precursor 1,3-diaminopropane, has been previously found in plants and was shown to originate in the enzymic

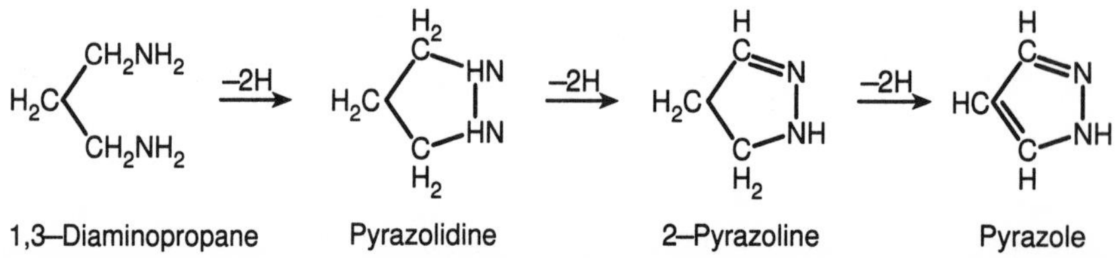

1,3–Diaminopropane —(−2H)→ Pyrazolidine —(−2H)→ 2–Pyrazoline —(−2H)→ Pyrazole

Fig. 3.2 Biosynthesis of pyrazole from 1,3-diaminopropane. The enzyme catalysing this series of dehydrogenations, pyrazole synthase, is FAD-dependent and 2-pyrazoline has been shown to be an intermediate in the process.

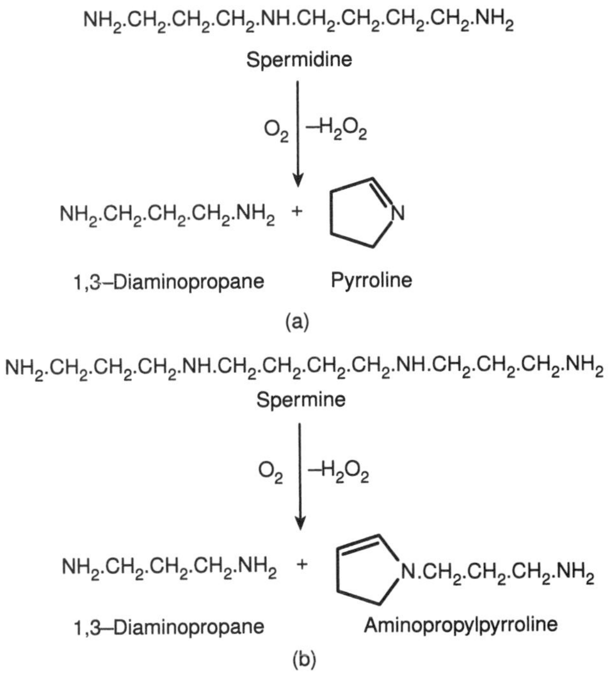

$NH_2.CH_2.CH_2.CH_2.NH.CH_2.CH_2.CH_2.CH_2.NH_2$

Spermidine

O_2 | $-H_2O_2$

$NH_2.CH_2.CH_2.CH_2.NH_2$ + N

1,3–Diaminopropane Pyrroline

(a)

$NH_2.CH_2.CH_2.CH_2.NH.CH_2.CH_2.CH_2.CH_2.NH.CH_2.CH_2.CH_2.NH_2$

Spermine

O_2 | $-H_2O_2$

$NH_2.CH_2.CH_2.CH_2.NH_2$ + $N.CH_2.CH_2.CH_2.NH_2$

1,3–Diaminopropane Aminopropylpyrroline

(b)

Fig. 3.3 1,3-Diaminopropane, the precursor of pyrazole, originates from the action of polyamine oxidase on (a) spermidine, and (b) spermine.

oxidation of spermidine and spermine by the enzyme polyamine oxidase (Fig. 3.3).

Availability of pyrazole is the limiting factor in the formation of β-pyrazol-1-ylalanine. The biosynthetic enzyme, β-pyrazolylalanine synthase, virtually scavenges for pyrazole in what appears to be a detoxication process (Brown, 1995). Accumulation of pyrazole is toxic to plant and animal cells whereas that of β-pyrazolylalanine is not. β-pyrazol-1-ylalanine synthase is a pyridoxal 5′-phosphate-dependent enzyme that uses *O*-acetylserine as the donor of the alanine residue transferred to pyrazole (Fig. 3.4).

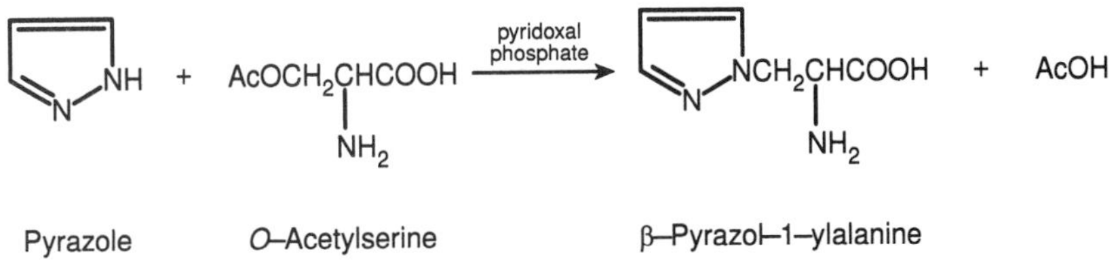

NH + $AcOCH_2CHCOOH$ (with NH_2) —pyridoxal phosphate→ $NCH_2CHCOOH$ (with NH_2) + AcOH

Pyrazole *O*–Acetylserine β–Pyrazol–1–ylalanine

Fig. 3.4 Formation of β-pyrazol-1-ylalanine from pyrazole and *O*-acetylserine. The reaction is catalysed by β-pyrazol-1-ylalanine synthase, a pyridoxal 5′-phosphate-dependent enzyme.

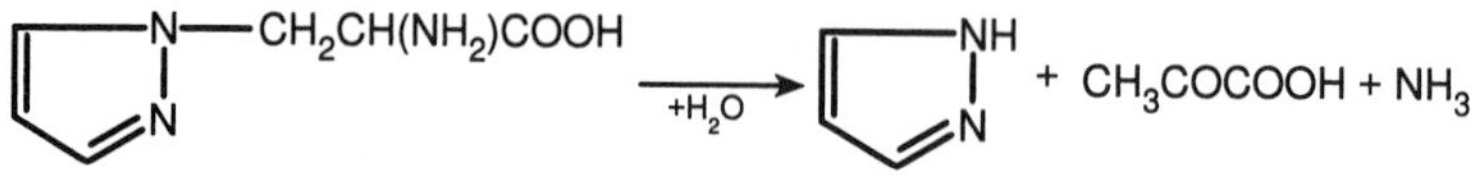

$$\text{pyrazol-1-yl-}CH_2CH(NH_2)COOH \xrightarrow{+H_2O} \text{pyrazole (NH)} + CH_3COCOOH + NH_3$$

Fig. 3.5 Microbial hydrolysis of β-pyrazol-1-ylalanine to yield free pyrazole, pyruvate and ammonia.

3.4 CATABOLISM

β-pyrazol-1-ylalanine is not catabolized to any detectable extent by mammalian tissues. In experiments in which [^{14}C]β-pyrazol-1-ylalanine was added to the diet of mice, more than 93% of the radioactivity was recovered in the urine as unmetabolized β-pyrazol-1-ylalanine. There was no evidence of the enzymic activity reportedly found in species of *Pseudomonas* (Takeshita *et al.*, 1963) and which catalyses the release from the amino acid of free pyrazole (Fig. 3.5). In view of the toxicity of pyrazole (Section 3.5) and the dietary occurrence of β-pyrazol-1-ylalamine, this is fortunate.

Pyrazole, administered subcutaneously to mice, is largely excreted in the urine as a peptide conjugate of 3,4,4-trimethyl-5-pyrazolone. The occurrence of this pyrazolone, identified by mass spectrometry and p.m.r. spectroscopy, implies that C-methylation occurs, a process not previously observed in a mammalian detoxication context.

The synthetic pyrazolinone drug antipyrine (2,3-dimethyl-1-phenyl-3-pyrazolin-5-one) is mainly oxidized in mammals to 4-hydroxyantipyrine but about 5% is eliminated unchanged in the urine. Another fraction is oxidized at the 3-methyl group, to give 3-hydroxymethyl-2-methyl-1-phenyl-3-pyrazolin-5-one.

3.5 BIOLOGICAL AND PHARMACOLOGICAL ACTIVITY

Neither pyrazole nor β-pyrazol-1-ylalanine has any known biological function but the former has marked pharmacological activity. As mentioned above, pyrazole is toxic in mammalian systems. It is a potent competitive inhibitor of liver alcohol dehydrogenase (Theorell and Yonetani, 1963; Theorell, Yonetani and Sjoberg, 1969; Li and Theorell, 1969) and inhibits the microsomal ethanol-oxidizing system. In addition, it inhibits the activity of the enzyme catalase and the transport into mitochondria of the biochemical reductants NADH and NADPH (Cederbaum and Rubin, 1974). Chronic administration of pyrazole induces ultrastructural changes in rat liver and affects brain noradrenalin concentrations. Pyrazole affects enzymes, such as NADP-cytochrome *c* reductase, that are involved in hydroxylation processes (Marselos *et al.*, 1977), and also those enzymes involved in glucuronidation. The activities of UDP-glucose dehydrogenase, UDP-glucuronyl transferase, and L-gulonate dehydrogenase, are markedly enhanced by administration of pyrazole, whereas the activities of UDP-glucuronic acid pyrophosphatase, β-glucuronidase, and D-glucuronolactone dehydrogenase are decreased (Marselos, 1977).

Pyrazomycin (5) and Formycin (6) are both antibiotics. Formycin, produced by *Streptomyces lavendulae*, is primarily active against *Xanthomonas oryzae*, the inhibition concentration being 6 $\mu g.mol^{-1}$. Pyrazomycin is obtained from *Streptomyces candidus* and shows antifungal activity, especially against *Neurospora*. It is active *in vivo* against Herpes simplex, the virus that causes 'cold sores', and also against vaccinia. It has been proposed for the clinical treatment of virus diseases at a dosage of 0.5–250 mg kg^{-1} given daily.

3.6 SYNTHETIC PYRAZOLES OF MEDICAL AND AGRICULTURAL IMPORTANCE

Antipyrine (7) was first synthesized, by Knorr, in 1884 and used as an antipyretic, later as an analgesic. It went out of favour clinically in the 1930s as new analgesics became available. There are, however, a number of pyrazole antipyretics, anti-inflammatory drugs and analgesics in present-day pharmacopoeias. Examples include Epirizole (8) and Difenamizole (9) listed as analgesic, antipyretic and anti-inflammatory, and Phenylbutazone (10) commonly used clinically as an anti-inflammatory in arthritis but which also has analgesic and antipyretic properties.

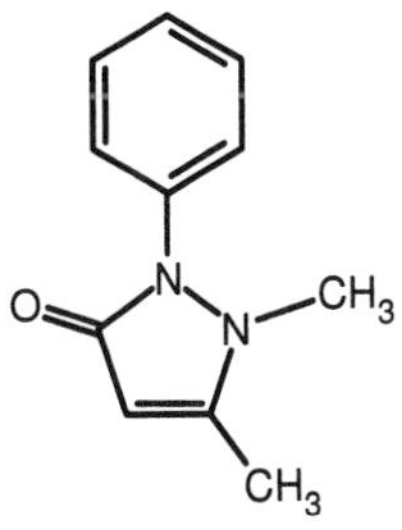

(7)

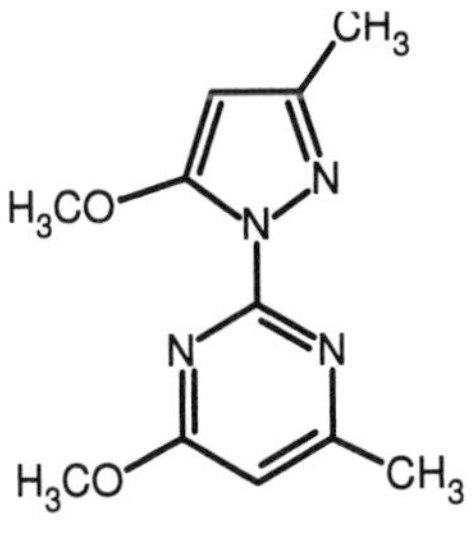

(8)

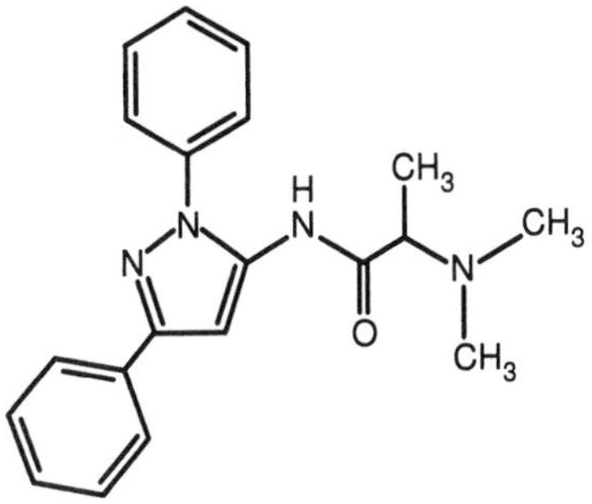

(9)

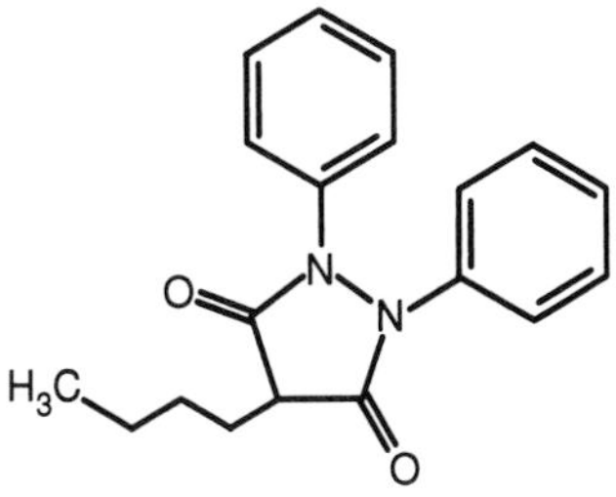

(10)

A synthetic aza-pyrazole, 1-β-D-ribofuranosyl-1*H*-1,2,4-triazole-3-carboxamide (11), known as Ribavirin or Virazole, is a powerful, clinically useful virostat with broad spectrum activity. it is the first synthetic non-interferon-inducing, broad-spectrum antiviral nucleoside and has been used with some success against HIV.

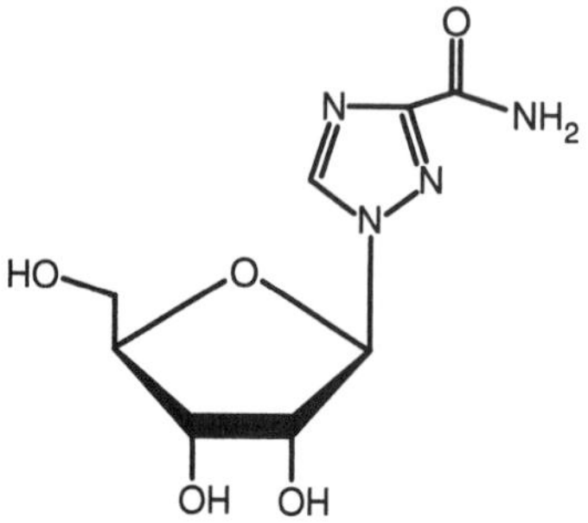

(11)

A number of pyrazole derivatives find application in agriculture and horticulture. Difenzoquat (12) is used as a post-emergence herbicide for food crops and has the added advantage of being an effective fungicide, especially against mildew. A group of pyrazolopyrimidine phosphate esters have been introduced in recent years as useful systemic fungicides. These compounds are exemplified by 2-(*O*,*O*-diethylthionophosphoryl)-5-methyl-6-ethoxy-carbonylpyrazolo-(1,5,a)-pyrimidine (13).

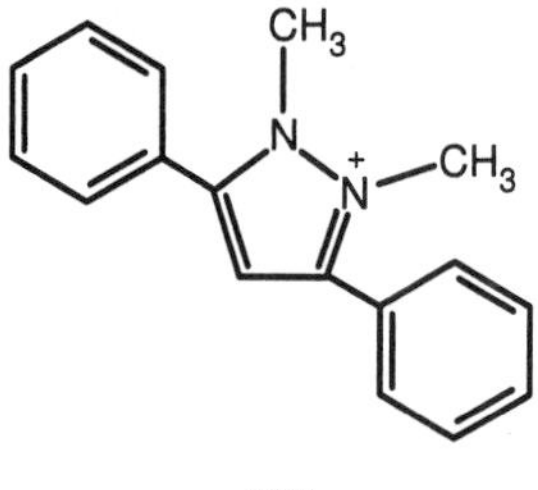

(12) 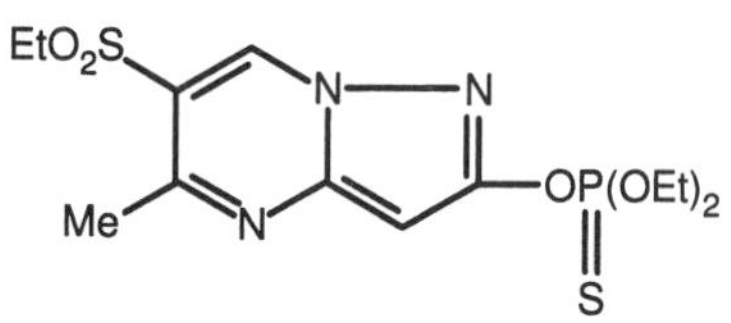 (13)

In addition to its clinical usefulness, described above, Ribavirin (11) has been used successfully in an agricultural context. Many commercially important stock strains of crop-plants, e.g. seed potatoes, contain inherent viral infections which whilst not sufficient to be regarded as overt diseases, do reduce the vigour of the crop. With potatoes and with tobacco, this problem has been shown to be eradicable by taking a sample of the chosen tissue into cell-suspension culture, treating it with Ribavirin, and then regenerating plantules from the virus-free cells.

REFERENCES

Brown, E.G. (1995) Biogenesis of *N*-heterocyclic amino acids by plants, in *Amino Acids and Their Derivatives*, (ed R.M. Wallsgrove), SEB Symposium Series Vol. 56, Cambridge University Press, Cambridge, pp. 119–144.

Brown, E.G. and Diffin, F.M. (1990) Biosynthesis and metabolism of pyrazole by *Cucumis sativus*: enzymic cyclization and dehydrogenation of 1,3-diaminopropane. *Phytochemistry*, **29**, 469–478.

Brown, E.G., Flayeh, K.A.M. and Gallon, J.R. (1982) The biosynthetic origin of the pyrazole moiety of β-pyrazol-1-yl-L-alanine. *Phytochemistry*, **21**, 863–867.

Cederbaum, A.I. and Rubin, E. (1974) Effects of pyrazole, 4-bromopyrazole and 4-methylpyrazole on mitochondrial function. *Biochemical Pharmacology*, **23**, 203–13.

Dunnill, P.M. and Fowden, L. (1963) The biosynthesis of β-pyrazol-1-ylalanine. *Jounral of Experimental Botany*, **14**, 237–248.

LaRue, T.A. (1965) Spectrometric determination of pyrazole with sodium amminoprusside. *Analytical Chemistry*, **37**, 246–248.

Li, T.K. and Theorell, H. (1969) Human liver alcohol dehydrogenase: inhibition by pyrazole and pyrazole analogues. *Acta Chemica Scandinavica*, **23**, 892–902.

Marselos, M., Törrönen, P., Alakuijala, P. and MacDonald, E. (1977). Hepatic hydroxylation and glucuronidation in the rat after subacute pyrazole treatment. *Toxicology*, **8**, 251–261.

Noe, F.F. and Fowden, L. (1960) β-pyrazol-1-ylalanine, an amino acid from water-melon seeds (*Citrullus vulgaris*). *Biochemical Journal*, **77**, 543–547.

Shinano, S. and Kaya, T. (1957). A new amino acid in water-melon juice. *Nippon Nôgei Kagaku Kaishi*, **31**, 759–762.

Takeshita, M., Nishizuka, Y. and Hayaishi, O. (1963) Studies on β-(pyrazol-*N*)-L-alanine. *Journal of Biological Chemistry*, **238**, 660–665.

Theorell, H. and Yonetani, T. (1963) Liver alcohol dehydrogenase–diphosphopyridine nucleotide (DPN)–pyrazole complex; a model of an intermediate in the enzyme reaction. *Biochemische Zeitschrift*, **338**, 537–553.

Theorell, H., Yonetani, T. and Sjoberg, B. (1969) Effects of some heterocyclic compounds on the activity of liver alcohol dehydrogenase. *Acta Chemica Scandinavica*, **23**, 255–260.

CHAPTER 4

Pyridines

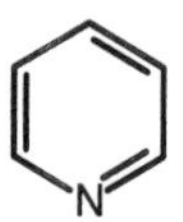

4.1 DISCOVERY AND NATURAL OCCURRENCE

Pyridine was discovered in the middle of the last century (Anderson, 1849) as a product of the dry distillation of bones. The first isolation of a pyridine derivative from living tissue was by Johns (1885) who obtained crystalline samples of trigonelline (*N*-methylpyridinium-3-carboxylic acid; 1) from the plant *Trigonella foenum graecum*. Biochemically however, the most important discovery concerning the natural occurrence and importance of pyridine derivatives can be traced back to the observations of Harden and Young (1904) that alcoholic fermentation by yeast requires a heat-stable, dialysable (i.e. low M_r) fraction to which the name cozymase was given. During the period 1935–36, Warburg and his collaborators identified a hydrogen-transferring coenzyme, from red blood cells, as a dinucleotide of adenine and nicotinamide. Von Euler's group then showed that cozymase was of a similar composition and only differed from the hydrogen transferring coenzyme by having 3 atoms of phosphorus per molecule, whereas the latter coenzyme contained 2 per molecule. These two redox coenzymes are what we now know as NAD (nicotinamide adenine dinucleotide) and NADP (nicotinamide adenine dinucleotide phosphate).

COOH

$\overset{+}{N}$

CH_3

(1)

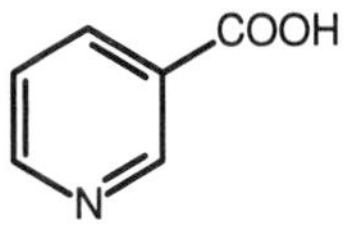

(2)

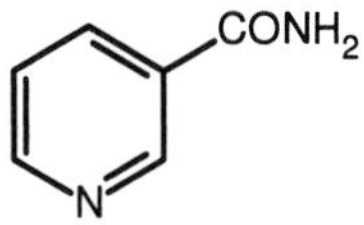

(3)

Nicotinic acid (pyridine 3-carboxylic acid; 2) the biological precursor of nicotinamide (3) was first isolated, simultaneously but independently, by Funk and by Suzuki in 1912. They had been investigating the nature of a water-soluble factor, present in rice bran, which could cure and prevent polyneuritis. At the time they did not recognize the dietary essentiality of this compound, now known to be a vitamin. When Funk and Suzuki made their discovery, the structure of nicotinic acid had already been known for 40 years as a laboratory isolate from the oxidation of nicotine. The realization that nicotinic acid is a vitamin came from the subsequent studies of Elvehjem who, in 1937, showed that the canine disease 'black-tongue' could be cured and prevented by adding nicotinic acid to the diet. During the period 1937–38, independent reports by Fonts, Smith, and Spies described the curing and prevention of pellagra in humans. Once prevalent in the poorer parts of the southern states of America where it was brought about by a nicotinic acid-deficient staple diet of maize, molasses and fat pork, pellagra is clinically characterized by 'the three ds', dermatitis, diarrhoea, and dementia.

Another water-soluble vitamin, pyridoxine (4), is also a substituted pyridine. It was first identified as a nutritional requirement for rats by Györgyi (1934). In these animals its deficiency leads to a characteristic dermatitis known as 'acrodynia' which can be cured by administering pyridoxine or either of the related compounds pyridoxal (5) and pyridoxamine (6). Pyridoxine functions biochemically, in the form of pyridoxal 5′-phosphate, as the coenzyme of a variety of amino acid-metabolizing enzymes, e.g. aminotransferases, amino acid decarboxylases, amino acid deaminases (Section 4.5).

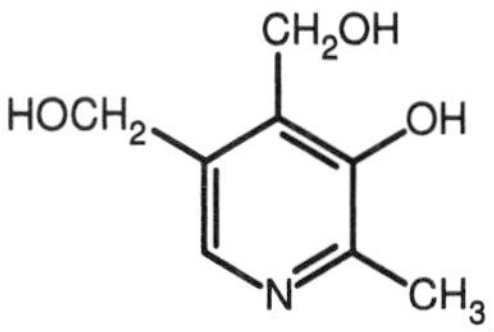

(4)

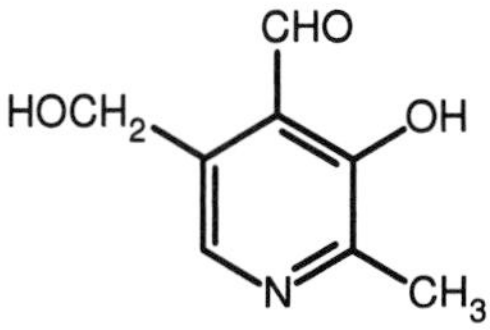

(5)

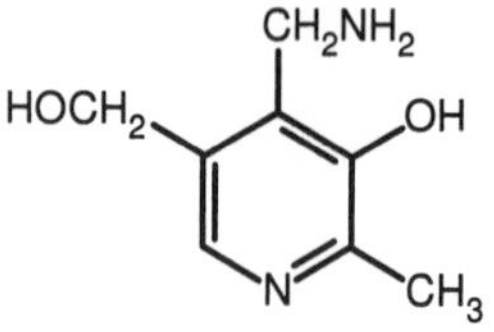

(6)

Simple substituted pyridines play an important role as flavour constituents of many foods and beverages. Table 4.1 gives examples of this function. Structurally more complex pyridine derivatives occur as alkaloids of which nicotine (7) is an example. Undoubtedly because of the commercial value of tobacco leaf and the extensive analyses that have been carried out on this product, a number of pyridine alkaloids have been identified from the same source. The more important of these, quantitatively, are listed in Table 4.2.

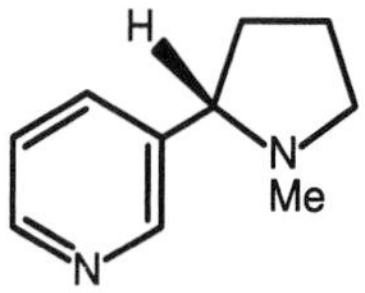

(7)

A pyridine natural product that appears both in lists of alkaloids and in those of unusual amino acids is mimosine (8). This pyridone, which has interesting biological activity (Section 4.6), takes its name from the plant family Mimosoideae in which it is widely distributed. Ricinine (3-cyano-4-methoxy-*N*-methylpyridin-2-one; 9) is an alkaloid produced by the castor bean plant (*Ricinus communis*). Fusaric acid (10), a systemic wilt toxin found particularly in cotton plants, is produced by various species of *Fusarium* and other fungi.

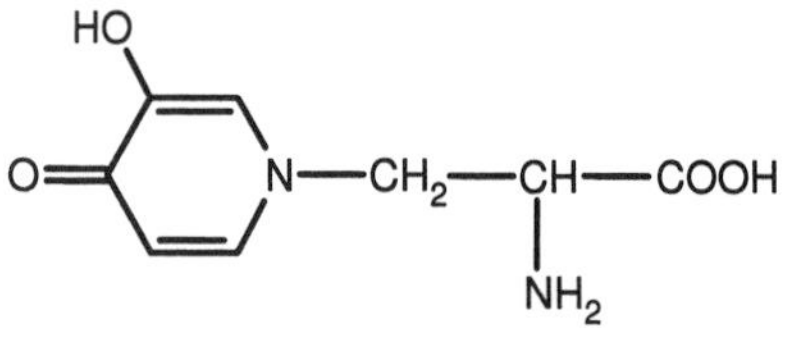

(8)

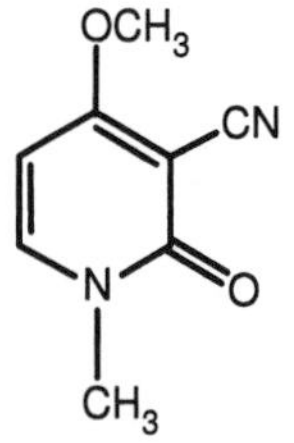

(9)

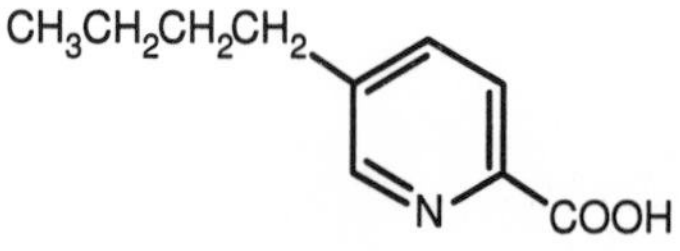

(10)

Table 4.1 Pyridines in food and drinks

Roast meat	**White bread**	**Black tea**
2-ethylpyridine	pyridine	pyridine
5-ethylpyridine	2-methylpyridine	picolines
2-hexylpentylpyridine	2-acetylpyridine	2-ethylpyridine
2-pentylpyridine	3-acetylpyridine	3-ethylpyridine
3-pentylpyridine	2-formylpyridine	2,6-dimethylpyridine
3,4-dimethylpyridine	2-methyl-5-ethylpyridine	2-methyl 5-ethylpyridine
Smoked fish	picolines	2-methyl 6-ethylpyridine
pyridine	4-ethylpyridine	3-methoxypyridine
Roasted nuts	2,3-dimethylpyridine	4-vinylpyridine
pyridine	**Roasted coffee**	2-acetylpyridine
2-methylpyridine	pyridine	2-phenylpyridine
2-pentylpyridine	**Roasted cocoa**	3-phenylpyridine
2-acetylpyridine	2-methylpyridine	**Beer**
methylnicotinate	3-vinylpyridine	2-acetylpyridine
	2-methyl-5-ethylpyridine	3-acetylpyridine
	2-acetylpyridine	**Rum and whisky**
	3-phenylpyridine	2-methylpyridine
		3-methylpyridine

Table 4.2 Pyridine alkaloids of tobacco leaf

N′-acetylnornicotine
anabasine
anatabine
anatalline
cotinine
N′-formylnornicotine
N′-hexanoylnornicotine
isonicoteine (2,3′-bipyridyl)
metanicotine
N′-methylanabasine
N′-methylanatabine
5-methyl-2,3′-bipyridyl
N′-methylnicotone
myosmine
nicotelline
nicotine
nicotyrine
N′-nitrosonicotine
nornicotine
nornicotyrine
N′-octanoylnornicotine
oxynicotine (nicotine *N*-oxide)
1,3,6,6-tetramethyl-5,6,7,8-tetrahydroisoquinolin-9-one
3,6,6-trimethyl-5,6-dihydro-7*H*-pyrindan-7-one

4.2 CHEMICAL PROPERTIES OF BIOCHEMICAL INTEREST

Although pyridine is a highly aromatic heterocycle, the heteroatom gives it a distinctly different character from that of benzene. For example, as the aromatic sextet of π-electrons is complete without the need to invoke the lone pair on the nitrogen, the latter is readily available to form bonds without disturbing ring aromaticity. There is, thus, a site for protonation, alkylation, and acylation which has no counterpart in benzene. Many of the chemical properties of pyridine are therefore those of a tertiary amine. It is weakly basic (pK_a 5.2) and can be quaternized with alkylating agents to form pyridinium salts. Because of the heteroatom, the pyridine ring is not very susceptible to normal electrophilic aromatic substitution but nucleophilic attack is facilitated, especially at the 2- and 4- positions.

Being π-deficient, pyridine is more easily reduced than benzene. Even at atmospheric pressure it can be fully reduced by hydrogen in the presence of Raney nickel to yield piperidine. Sodium/ethanol or lithium aluminium hydride will also reduce pyridine, albeit not fully, to give tetrahydropyridine. Alkyl pyridinium salts are more electrophilic than pyridine and can be partially reduced even by the mild reductant sodium borohydride. This facile reduction of pyridinium salts is of substantial importance in relation to biochemical redox reactions involving NAD^+ and $NADP^+$ (Section 4.5).

Of practical importance in the analytical laboratory is the marked instability of NAD^+ and $NADP^+$ in alkaline solutions, and that of NADH and NADPH in dilute acid. The alkali-instability results from the ease with which the pyridine moiety forms adducts with OH^-. Adducts of OH^- in the 2 or 4 position of the pyridine ring undergo facile ring opening (Fig. 4.1) and can be followed by further degradation. When heated at 100° in dilute alkali, NAD^+ is rapidly hydrolysed to nicotinamide and ADP-ribose.

In mildly acidic solutions, the pyridine moiety of NADH and NADPH is protonated at C-5 and then OH^- adds at the 6-position

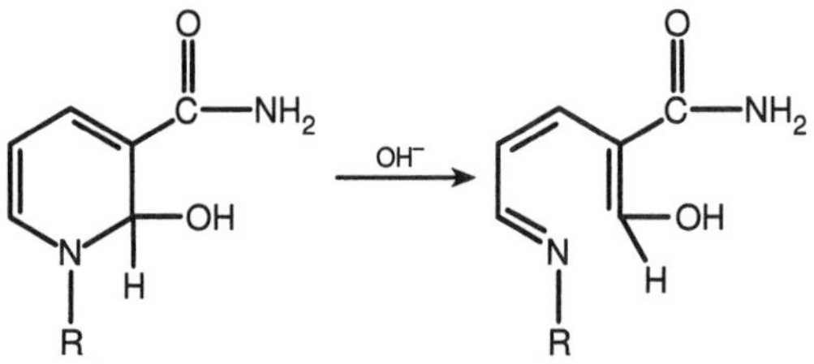

Fig. 4.1 In alkaline solution, NAD^+ forms adducts with OH^- at either the 2 or 4 position. These readily ring-open, as illustrated above for the 2-adduct, accounting for the alkaline instability of this coenzyme.

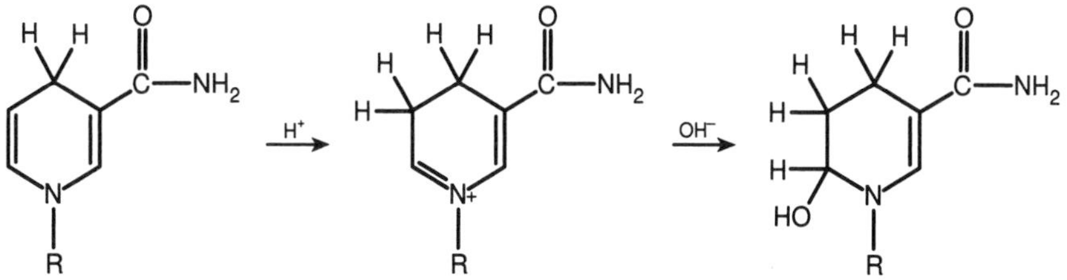

Fig. 4.2 Reduced pyridine nucleotides are rapidly degraded in dilute acidic solution. This is because the ring becomes protonated at C-5 and then a nucleophile, such as OH^-, adds at C-6 as shown. The adduct can be further broken down by water adding aross the other double bond and the ring opening either side of the *N*-atom.

(Fig. 4.2). Water can also add across the other double bond and the compound then undergoes ring-opening on either side of the *N*-atom (Fig. 4.2).

4.3 BIOSYNTHESIS AND INTERCONVERSION

The pyridine ring of nicotinamide adenine dinucleotide (NAD) and that of nicotinamide adenine dinucleotide phosphate (NADP), Fig. 4.3, both

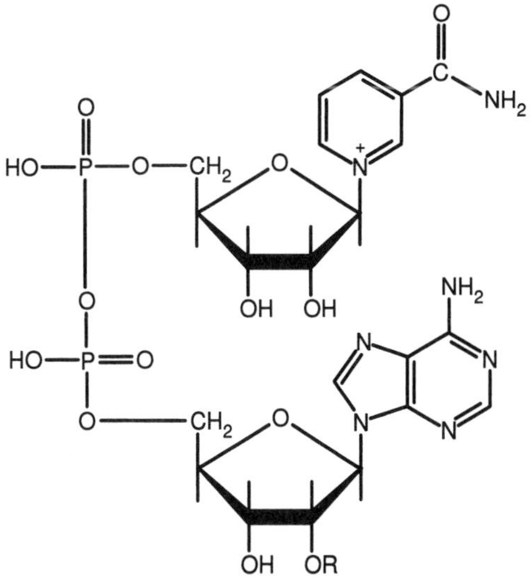

Fig. 4.3 Structure of the oxidized form of nicotinamide adenine dinucleotide (NAD^+) in which R=H; and the oxidized form of nicotinamide adenine dinucleotide phosphate ($NADP^+$) in which R is an orthophosphate group. The additional phosphate group on $NADP^+$ is located at the 2′-position of the adenosine moiety.

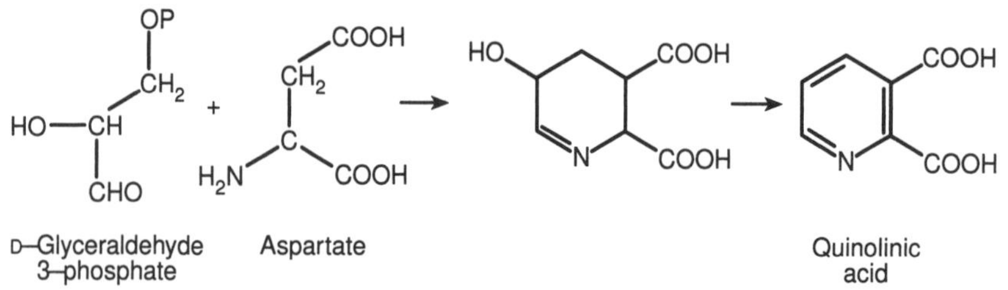

D–Glyceraldehyde 3–phosphate + Aspartate → → Quinolinic acid

Fig. 4.4 Quinolinic acid, the precursor of the pyridine ring of NAD and NADP, is synthesized in plants from aspartate and glyceraldehyde phosphate.

arise *de novo* from quinolinic acid. In plants this is mainly formed from aspartate and glyceraldehyde phosphate (Fig. 4.4) but in animal tissues and microorganisms it is supplied by the catabolism of tryptophan (Fig. 4.5). Thus, providing there is a plentiful supply of tryptophan, animals can meet their requirements. When however dietary intake of tryptophan is so low as to become limiting for NAD and NADP synthesis, nicotinic acid assumes a vitamin role and an exogenous supply is therefore needed in these circumstances. Dietary nicotinic acid is enzymically *N*-phosphoribosylated by 5′-phosphoribosyl-1-pyrophosphate (PRPP) to nicotinic acid ribonucleotide and enters the biosynthetic scheme shown in Fig. 4.6.

In plants, the pyridine alkaloid ricinine (9) is synthesized from nicotinic acid *via* *N*-methyl-3-cyanopyridine. It is probable that the biosynthesis proceeds at the nucleotide level with the ribose 5-phosphate side chain being displaced by methyl at a late stage in the pathway. Nicotine is synthesized in tobacco plants by the condensation of dihydronicotinic acid with the *N*-methyl Δ^1-pyrrolinium cation. Similar condensation with Δ^1-piperideine yields the related alkaloid anabasine (11).

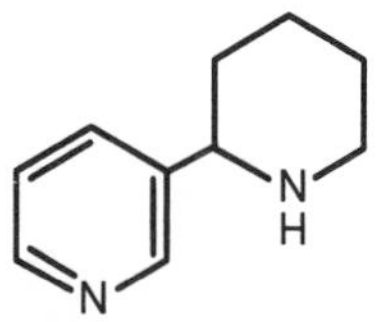

(11)

Trigonelline (1), a widely distributed pyridine alkaloid in plants and also found in animals, is formed by direct methylation of nicotinic acid. *N*-Methylnicotinamide, the corresponding carboxamide, is also found in animal tissue and often occurs as a constituent of urine. It originates in the catabolism of NAD^+ and $NADP^+$.

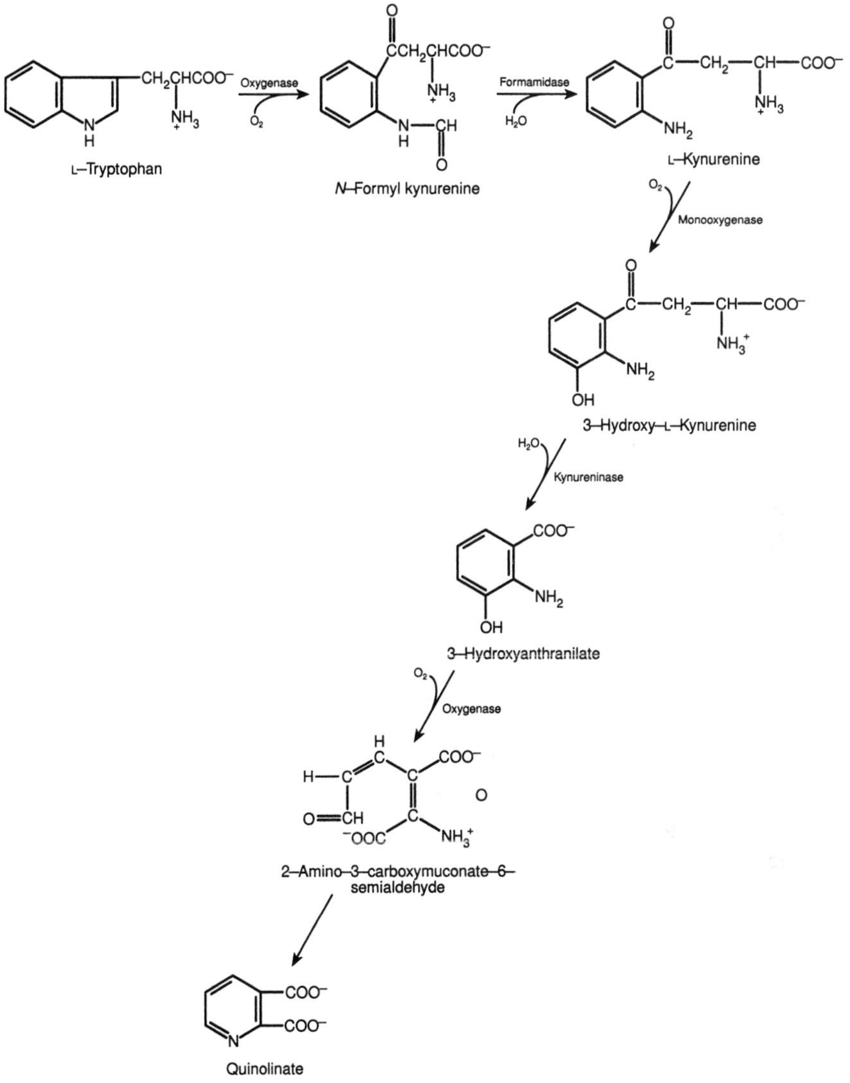

Fig. 4.5 Catabolic pathway for tryptophan, leading to the formation of quinolinic acid in animal tissues.

4.4 CATABOLISM

NAD^+ can be hydrolysed by either NAD-glycohydrolase or by NAD^+-pyrophosphatase (Fig. 4.7). The former releases nicotinamide and ADP-ribose whereas the latter produces nicotinamide mononucleotide (NMN) and AMP. Nicotinamide mononucleotide is further hydrolysed

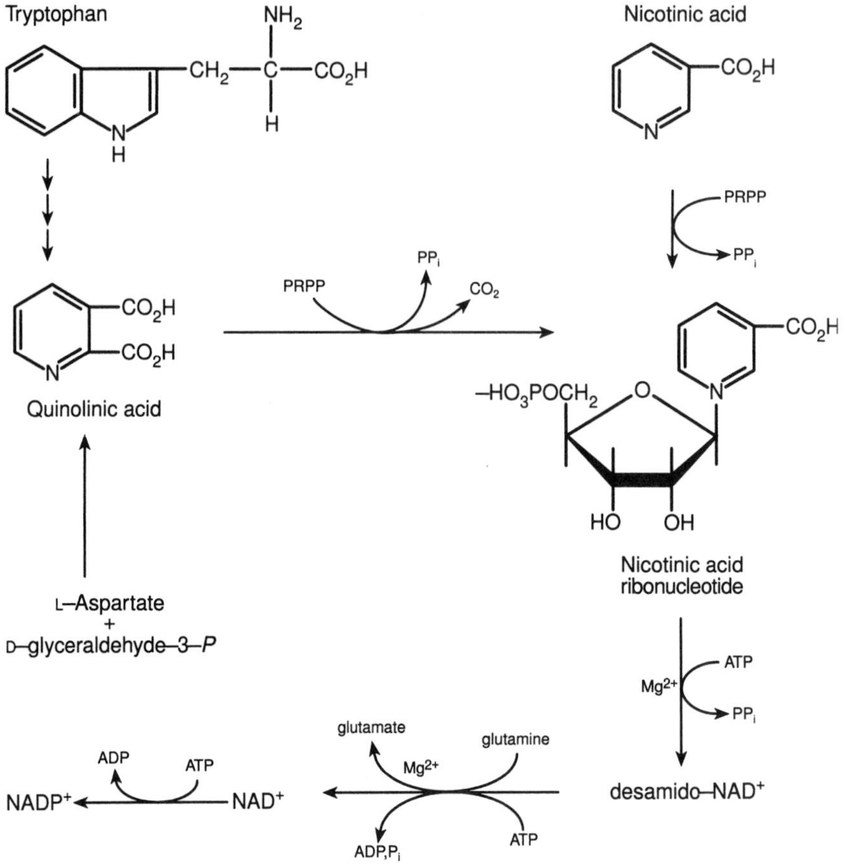

Fig. 4.6 Biosynthesis of the pyridine nucleotides, NAD^+ and $NADP^+$, which arise from tryptophan (Fig. 4.5) or dietary nicotinic acid in animal tissues, and from aspartate and glyceraldehyde 3-phosphate in plants.

to nicotinamide and ribose-1-phosphate. The nicotinamide from both of these sources is methylated and then oxidized (Fig. 4.7) and the two principal catabolites *N*-methylnicotinamide and 2-pyridone-5-carboxamide are excreted in the urine. When relatively large amounts of nicotinamide are being catabolised, these two catabolites impart a distinctive bright yellow fluorescence to the urine.

As discussed in Section 4.5 both cholera toxin and diphtheria toxin catalyse the transfer of an ADP-ribose residue from NAD^+ to recipient proteins or peptides. This, too, represents a significant catabolic route for NAD^+ in patients with these diseases.

The alcohol pyridoxine (4) and its corresponding aldehyde pyridoxal (5) are both catabolized by further oxidation to 4-pyridoxic

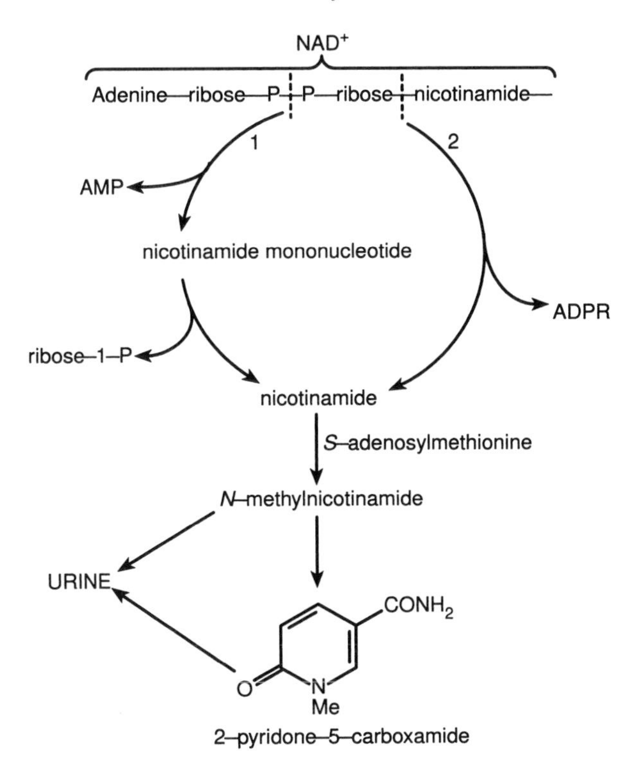

Fig. 4.7 NAD^+ can be hydrolysed by NAD^+-glycohydrolase (1) or by NAD^+-pyrophosphatase (2). The glycohydrolase releases nicotinamide and ADP-ribose (ADPR) whereas the pyrophosphatase produces nicotinamide mononucleotide (NMN) and AMP. The main catabolic products from NAD^+ are *N*-methylnicotinamide and 2-pyridone-5-carboxamide. When produced in excess, both compounds impart a distinctive yellow fluorescence to the urine.

acid (12) which is the principal urinary excretory product from this vitamin B_6 group of compounds. The oxidation takes place in the liver through the enzymic agency of an aldehyde oxidase.

COOH, $HOCH_2$, OH, N, CH_3

(12)

4.5 BIOCHEMICAL FUNCTIONS

As with most members of the B group of water-soluble vitamins, the pyridine derivatives nicotinic acid and pyridoxine function as compo-

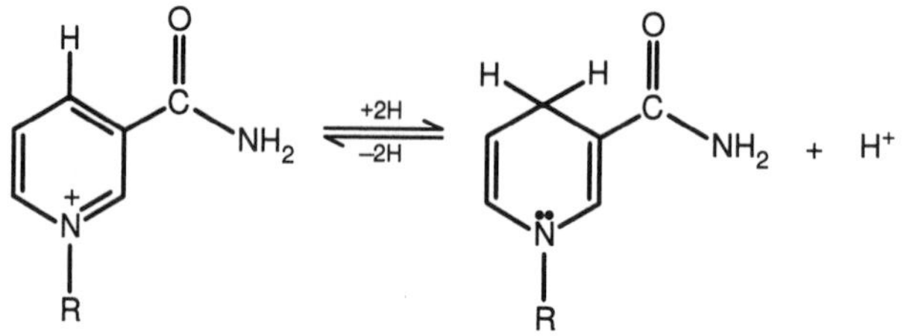

Fig. 4.8 Reversible reduction-oxidation of the coenzymes NAD^+ and $NADP^+$.

nents of coenzymes. Nicotinamide, the carboxamide of nicotinic acid forms the functional end of the NAD^+ molecule, whereas pyridoxine is the dietary precursor of pyridoxal 5′-phosphate, an important coenzyme in amino acid metabolism.

NAD^+ and its phosphate $NADP^+$ (Fig. 4.3) are the coenzymes of a majority of the dehydrogenases and readily undergo reduction to NADH and NADPH, respectively (Fig. 4.8). As can be seen from Fig. 4.8, although dehydrogenases remove 2 hydrogen atoms per molecule of oxidizable substrate, only one hydrogen atom and one electron are accepted by the coenzyme, and a proton is released into the surrounding medium. The reaction is reversed, i.e. NADH is reoxidized to NAD^+, when the reduced coenzyme in turn reduces an acceptor with a more positive redox potential, such as FAD (flavin adenine dinucleotide).

Experiments using deuterium-labelled substrates revealed that the enzymic reduction of NAD^+, and of $NADP^+$, by the dehydrogenases is stereospecific. As is shown in Fig. 4.9, some dehydrogenases, e.g. alcohol dehydrogenase and lactate dehydrogenase, transfer hydrogen to one side of the pyridine ring (the *pro-R* position) whereas others, such as glutamate dehydrogenase and glyceraldehyde 3-phosphate dehydrogenase, add hydrogen to the opposite side (*pro-S* position).

When the NAD^+ molecule is reduced, the pyridine ring loses its aromaticity (Fig. 4.8) and there is a change in spectrophotometric properties. Both the adenine and the nicotinamide chromophores absorb maximally at around 260 nm giving NAD^+ a strong absorption band in this region. Loss of the nicotinamide chromophore of NAD^+ by reduction causes a small decrease in the molar absorption coefficient at 260 nm but of more practical importance is the appearance of a new peak at 340 nm (Fig. 4.10).

This means that most biological dehydrogenations can be followed spectrophotometrically by the increasing absorption at 340 nm as the reaction proceeds. Extensive use is made of this in biochemistry, especially in clinical laboratories where changes in specific dehydrogenase activities can be of diagnostic value. For example, elevation of the activ-

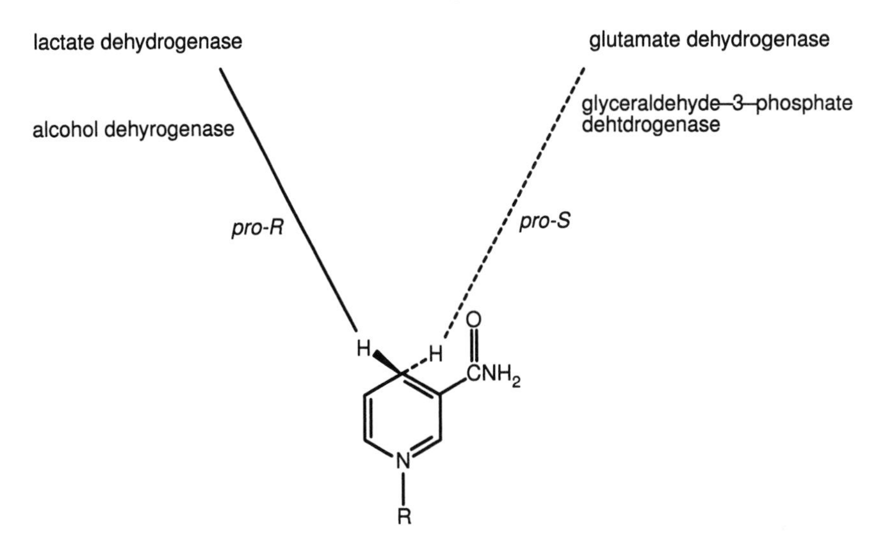

Fig. 4.9 The enzymic reduction of NAD^+ and $NADP^+$ is stereospecific. Some dehydrogenases transfer hydrogen to one side of the pyridine ring of the nicotinamide unit, others to the opposite side (*pro-R* and *pro-S*). Two examples of each type of dehydrogenase are given.

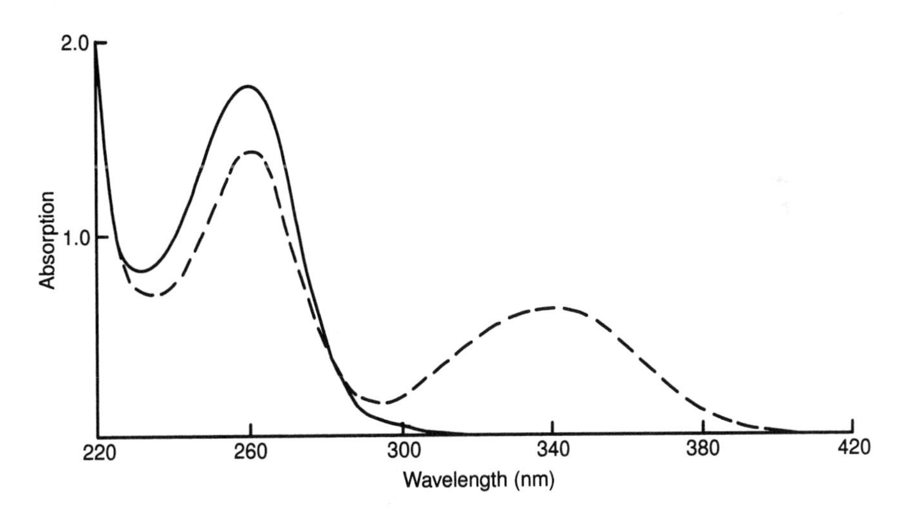

Fig. 4.10 Ultraviolet absorption spectrum of NAD^+ (solid line) and NADH (broken line) under physiological conditions of pH (pH 6–7).

ity of serum lactate dehydrogenase (LDH), especially of the isoenzymes LDH 1 and LDH 2, is associated with myocardial infarction.

As mentioned earlier (Section 4.4) NAD^+ also functions as a donor of ADP-ribose. Diphtheria toxin, produced by *Corynebacterium diphtheriae*, is a protein with a catalytic domain that is capable of using NAD^+ to ADP-ribosylate 'elongation factor 2'. The latter is a specific protein involved in protein synthesis within cells and its ADP-ribosylation leads to a cessation of protein production with fatal results for the individual cell, and eventually for the patient. Similarly using NAD^+, cholera toxin from *Vibrio cholerae* catalyses the ADP-ribosylation of the α-subunit of the G-protein ($G_{s\alpha}$) which normally brings about hydrolysis of GTP to GDP in the cyclic AMP-signalling system (Section 6.6). As a result, adenylyl cyclase becomes permanently activated and cyclic AMP concentrations rise causing a persistent and metabolically disruptive cyclic AMP signal. This is analogous to having a switch jammed in the 'on' position. One of the effects of this is to activate ion-pumps causing a large efflux of Na^+ and water into the gut, producing the characteristic and frequently lethal diarrhoea of cholera with its attendant dehydration and salt loss.

The second pyridine-derived vitamin, pyridoxine (4), also serves as a coenzyme precursor. The coenzyme in this case is pyridoxal 5′-phosphate (13). It serves as the coenzyme of virtually all enzymes that catalyse reactions involving amino acids. These reactions include transamination, racemization, α,β-elimination, β,γ-elimination, deamination, decarboxylation, and synthase reactions. Some examples are illustrated in Table 4.3. In each case, the reactant amino acid spontaneously and rapidly forms a Schiff base with pyridoxal 5′-phosphate (Fig. 4.11).

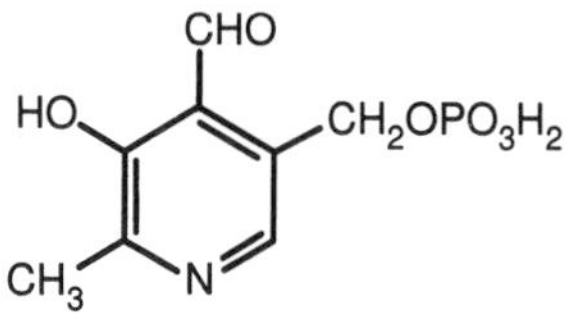

CHO
HO
$CH_2OPO_3H_2$
CH_3
N

(13)

Although the reactions mentioned above and exemplified in Table 4.3 appear, on the surface, to be quite different, they are made possible by this common structural feature. The Schiff's base provides a conjugated system of double bonds extending from the reaction site to the strongly electrophilic pyridinium nitrogen atom. Electron withdrawal occurs from the α-carbon atom of the attached amino acid towards the cationic nitrogen of the imine and into the electron sink afforded by the pyridoxal ring (Fig. 4.11). All three of the substituents on the α-carbon are thus labilized for reactions which require electron withdrawal from this atom. There are some reactions which require activation of the β-carbon by the imine nitrogen atom and this proceeds similarly.

In pyridoxal 5′-phosphate-dependent enzymic reactions, there is loss of a proton from the α-carbon atom of the amino acid moiety of the

Table 4.3 Pyridoxal phosphate dependent enzymic reactions

Type of enzyme	Examples of reactions catalysed
Racemase	$CH_3-CH(NH_2)-CO_2H \rightleftharpoons CH_3-CH(NH_2)-CO_2H$
Aminotransferase	$HO_2C-CH_2-CH_2-CH(NH_2)-CO_2H + HO_2C-CH_2-C(=O)-CO_2H \rightleftharpoons HO_2C-CH_2-CH_2-C(=O)-CO_2H + HO_2C-CH_2-CH(NH_2)-CO_2H$
Deaminase	$HO-CH_2-CH(NH_2)-CO_2H \xrightarrow{-H_2O} CH_2=C(NH_2)-CO_2H \xrightarrow{H_2O,\ -NH_3} CH_3-C(=O)-CO_2H$
Decarboxylase	imidazolyl$-CH_2-CH(NH_2)-CO_2H \rightarrow$ imidazolyl$-CH_2-CH_2-NH_2 + CO_2$
Tryptophan synthase (β–subunit)	indole $+ HO-CH_2-CH(NH_2)-CO_2H \rightarrow$ indolyl$-CH_2-CH(NH_2)-CO_2H + H_2O$
Cysteine desulphydratase	$HS-CH_2-CH_2-CH(NH_2)-CO_2H \xrightarrow{-H_2S} CH_3-CH=C(NH_2)-CO_2H \xrightarrow{H_2O,\ -NH_3} CH_3-CH_2-C(=O)-CO_2H$

Schiff's base (Fig. 4.11) and this leads to formation of a resonance-stabilized carbanion which is the key intermediate in a further series of reactions. For example, addition of a proton back to the same position but from the opposite side, effects racemization. Addition of a proton to the carbonyl carbon of the pyridoxal moiety gives the Schiff's base of pyridoxamine (6) and an α-keto acid. The Schiff's base can then hydrolyse to release the keto acid, so affording deamination of the amino acid. Alternatively, reversing this process with a different α-keto acid regenerates pyridoxal phosphate and yields a new amino acid, so bringing about transamination.

That the foregoing interpretation of the mode of action of pyridoxal

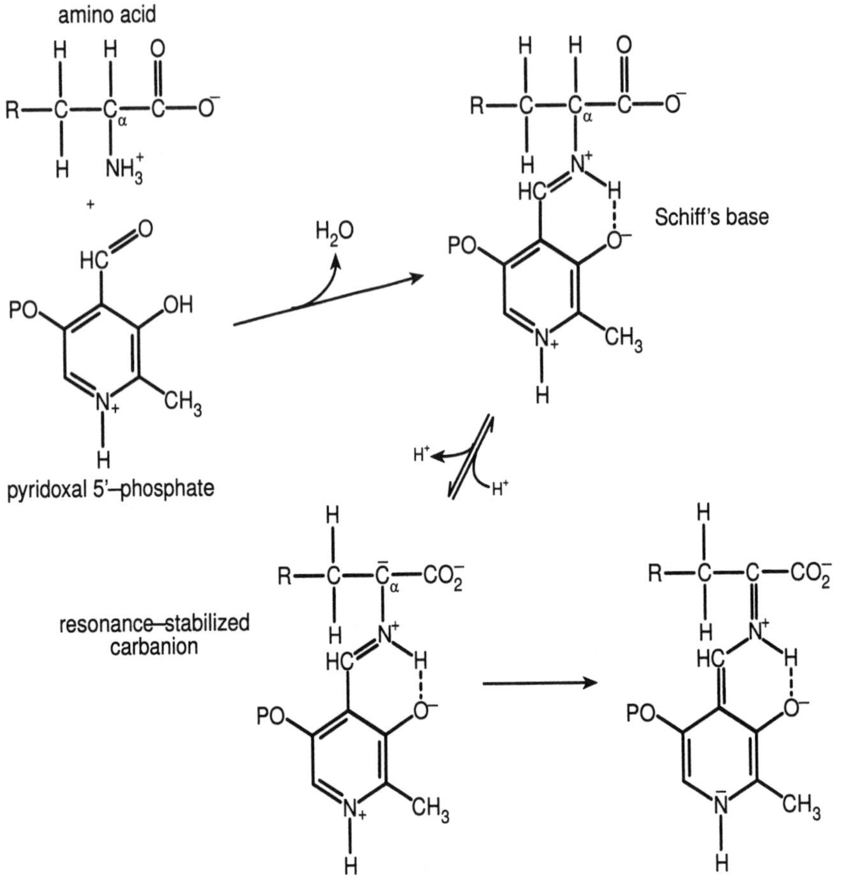

Fig. 4.11 An amino acid substrate reacts with the pyridoxal 5′-phosphate coenzyme, at the catalytic site of the enzyme, to form a Schiff's base. Withdrawal of electrons towards the imine nitrogen atom and into the electron sink provided by the pyridoxal ring, labilizes the three groups attached to the α-carbon atom of the amino acid. Loss of a proton, from the α-carbon of the amino acid moiety, then leads to formation of a resonance-stabilized carbanion which is the key intermediate in a further series of enzyme-catalysed reactions.

5′-phosphate is correct was amply supported by the work of Snell and his coworkers who duplicated many of the reactions using model, non-enzymic, systems (Metzler *et al.*, 1953). In these systems, the reactions are catalysed, in buffered solutions at various pH values, by appropriate divalent or trivalent metal cations at elevated temperatures (commonly 60–100°C); a Schiff's base-chelate (14) is involved. The choice of metal ion, pH and temperature, depends on the type of enzymic reaction being simulated.

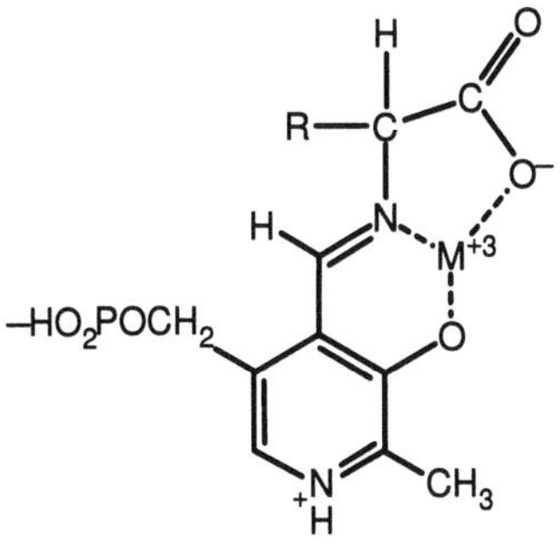

(14)

4.6 BIOLOGICAL AND PHARMACOLOGICAL ACTIVITY OF PYRIDINE DERIVATIVES

Pyridine itself, in traces, is a natural flavour constituent of a number of foodstuffs and beverages (Table 4.1). In these quantities, it appears to be harmless. In more significant amounts, however, it can cause depression of the central nervous system, and irritation, both of the skin and the respiratory tract. Larger amounts cause damage to the gastrointestinal tract, and to kidney and liver.

There are a number of simple pyridine and piperidine alkaloids with significant biological or pharmacological properties. Probably the best known of the pyridine alkaloids is nicotine (7). A typical filter cigarette contains 20–30 mg of nicotine, of which about 10% is absorbed by inhalation. Nicotine mimics acetylcholine and thereby stimulates neurotransmission at the cholinergic receptors and results in peripheral vasoconstriction and a rise in blood pressure. Since nicotine is not rapidly deactivated, it prevents incoming nerve impulses from being effective and so partially blocks the transmission of new information. In addition, nicotine activates adrenergic response by stimulating the release of adrenalin from the adrenal glands. It also blocks some heat and pain sensory receptors, such as those of the skin and tongue. In acute nicotine poisoning, tremors and convulsions are seen and the patient dies of respiratory paralysis. A well known pharmacological effect of nicotine is suppression of hunger contractions by the stomach and consequent loss of appetite. The piperidine alkaloid lobeline (15)

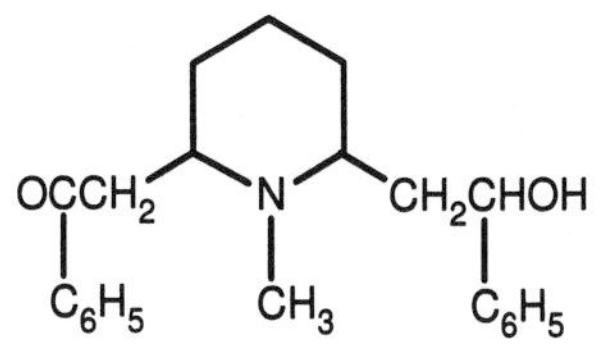

(15)

from *Lobelia inflata* is commonly used in preparations aimed at helping people to give up smoking since it has similar physiological effects to nicotine. Coniine (16) another piperidine alkaloid is the poisonous principal of hemlock (*Conium maculatum*). Hemlock poisoning causes vomiting and diarrhoea, inflammation of the gastrointestinal tract, mental confusion, convulsion and death.

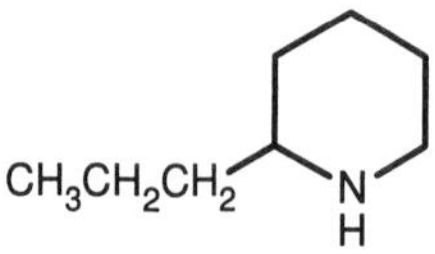

$CH_3CH_2CH_2$ NH

(16)

As both nicotinic acid and pyridoxine are vitamins they can be said to have biological activity but the effect is, of course, only really seen in treating cases of the corresponding vitamin deficiency. As is true of all vitamin deficiencies, the response to treatment with these compounds, or their precursors, is dramatic and a normal physiological state is quickly resumed. Both nicotinic acid and nicotinamide are now of interest in other therapeutic fields. In schizophrenia, for example, a number of patients have been found to be deficient in nicotinic acid and its amide, and it has been suggested that this contributes to the schizophrenic syndrome. There is clinical evidence of a positive response to megadoses of nicotinic acid but other vitamin supplements may also be required, e.g. pyridoxine supplements are claimed to facilitate endogenous production of nicotinic acid.

There are reports that nicotinamide inhibits poly(ADP-ribose) synthase and that at high concentrations it can act as a free radical scavenger. These and other observations have led to a claim that nicotinamide may also play a role in autoimmune diseases like rheumatoid arthritis, chronic colitis, and glomerulonephritis.

A number of poisons as well as useful drugs react with pyridoxal phosphate-requiring enzymes. For example, much of the toxic effect of carbonyl reagents, such as hydroxylamine, hydrazine, and semicarbazide arises from their ability to form stable derivatives, analogous to Schiff's bases, with pyridoxal phosphate. Isonicotinic acid hydrazide, known clinically as Isoniazid, INAH, INH and Rimifon, and by a variety of other proprietary names, has been one of the most effective drugs against tuberculosis since the early 1950s. The drug reacts with pyridoxal to form a hydrazone which inhibits pyridoxal kinase. This enzyme catalyses the phosphorylation of pyridoxal to pyridoxal 5′-phosphate, and as its activity in mycobacteria is low but nevertheless essential, the effect is to block the growth of these pathogenic microorganisms.

The pyridone amino acid mimosine (Section 4.1) also acts as a pyridoxal phosphate antagonist. This characteristic secondary product of the plant family Mimosoideae, to which belong mimosa and acacia, is known to be toxic in quantity to livestock, and to be teratogenic in mice and pigs. Interestingly, in addition to its antagonism of pyridoxal phosphate, mimosine blocks the synthesis of hair and wool protein and causes loss of hair in laboratory animals and in horses. Sheep that have ingested mimosine have been observed to shed their fleeces and this has led to an evaluation in Australia of the potential of mimosine as a chemical shearing agent.

Of agricultural and horticultural interest, fusaric acid (10) is a phytotoxin produced by various plant pathogenic species of the fungus *Fusarium*. It is a systemic poison, i.e. it is circulated internally within the infected plant, and affects the vascular tissue, causing the water supply to be cut off from various tissues and manifesting its presence by a wilting of the plant. Commercial crops affected in this way include cotton and tomato.

4.7 SYNTHETIC COMPOUNDS OF CLINICAL IMPORTANCE

A number of synthetic pyridines have been developed as pharmaceuticals. These include an anti-inflammatory compound Piroxicam (17) which is also known by a variety of other proprietary names. Nifedipine (18) is an anti-anginal, as is also Amlodipine (19); and Pinacidil (20) is a clinically useful hypertensive agent.

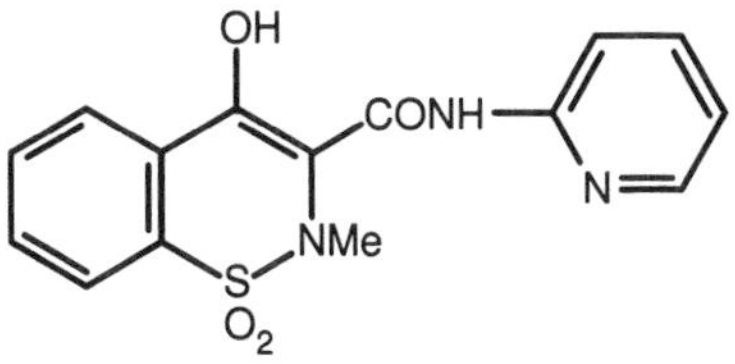

OH CONH N NMe S O_2

(17)

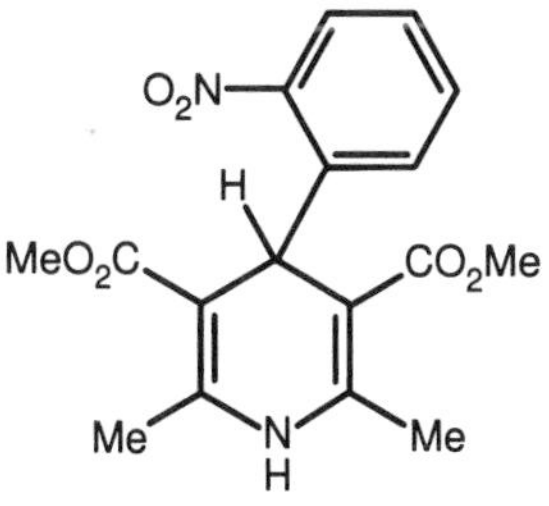

O_2N H MeO_2C CO_2Me Me N H Me

(18)

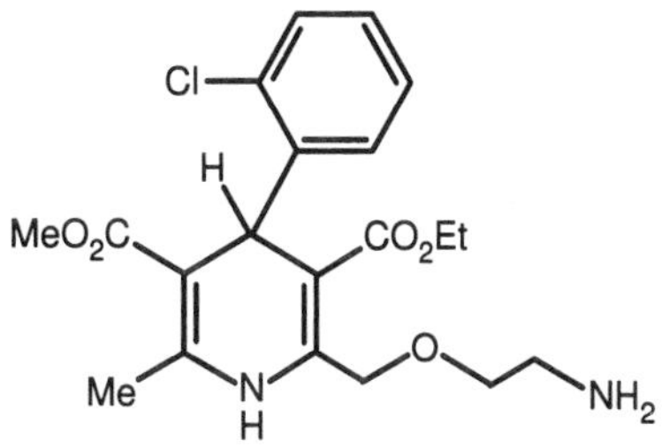

Cl H MeO_2C CO_2Et O Me N H NH_2

(19)

NCN Me HN N H CMe_3 N

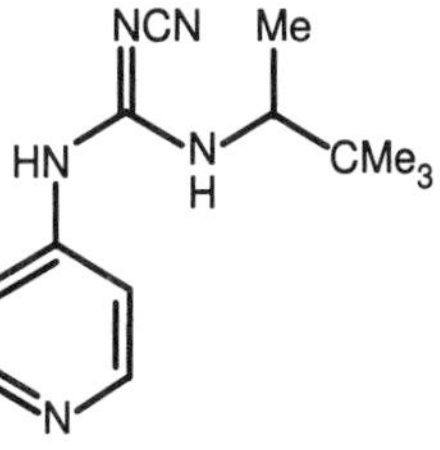

(20)

4.8 COMPOUNDS OF AGRICULTURAL IMPORTANCE

A major part of manufactured pyridine goes into the production of the well-known bipyridyl weedkillers diquat (21) and paraquat (22). Paraquat is, in effect, a methyl viologen, i.e. a redox compound (E'_0-0.44 V).

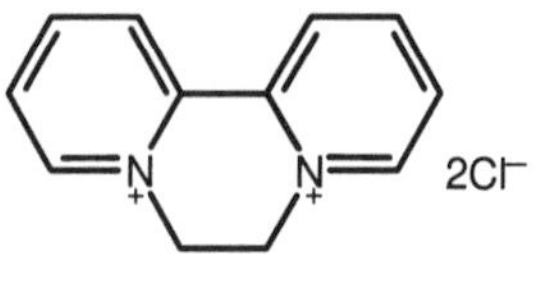

2Cl⁻

(21)

MeN⁺ N⁺Me 2Cl⁻

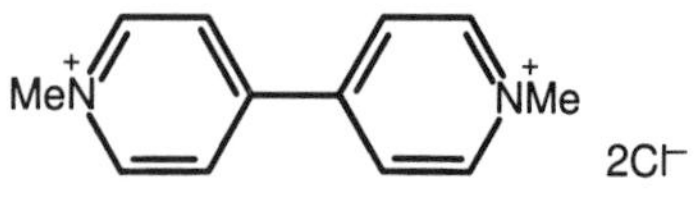

(22)

It will accept electrons from 'photosystem 1' in photosynthesis (Chapter 1, Section 1.8.3) at a position preceding the ferredoxins and so diverts the phytotoxic electron flow away from the reduction of $NADP^+$. The phytotoxic effect is, however, too rapid to be attributable solely to depletion of NADPH. It appears that the planar free radical produced when paraquat accepts an electron, reduces O_2 to the superoxide anionic free radical $O_2^{\cdot}$. The chloroplast enzyme superoxide dismutase converts this into hydrogen peroxide and O_2,

$$O_2^{\cdot} + O_2^{\cdot} + 2H^+ \rightarrow H_2O_2 + O_2$$

The H_2O_2 so produced oxidizes unsaturated fatty acyl residues in the chloroplastic membranes and inactivates them, preventing photosynthesis from taking place. In addition to its powerful action against photosynthetic organisms, paraquat is very toxic to animals and there have been several human fatalities caused by its accidental ingestion. Death occurs within 2–3 weeks from pulmonary fibrosis coupled with kidney and liver damage.

Chloropyridines are another group of commercially important herbicides. This includes Picloram (4-amino-3,5,6-trichloropicolinic acid; 23) which is highly active against broad leaved plants and conifers. Another, Lontrel (3,6-dichloropicolinic acid; 24) is tolerated by crops such as sugar beet and small grain cereals whilst exhibiting strong post-emergence activity against many of the weeds that are tolerant of phenoxyalkane carboxylic acid herbicides.

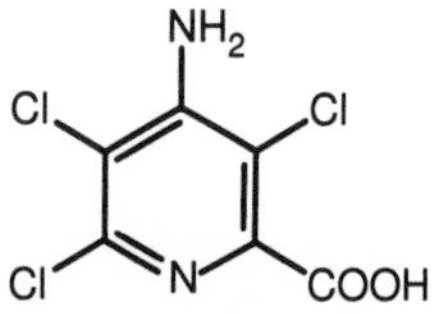

NH_2 Cl Cl Cl N COOH

(23)

Cl Cl N COOH

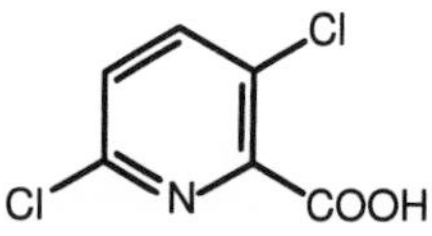

(24)

The pyridine ring also features prominently in insecticidal chemistry. For example, the pyridine derivatives chloropyrifos-methyl and -ethyl are examples of patented organothiophosphate insecticides which, like many organophosphate pesticides, function as cholinesterase inhibitors and exhibit broad insecticidal activity with low mammalian toxicity. Interestingly, a number of these cholinesterase inhibitions can be reversed by acid-salts of pyridine 2-carbaldehyde oxime (25).

N CH=NOH

(25)

REFERENCES

Metzler, D.E., Ikawa, M. and Snell, E.E. (1953) A general mechanism for vitamin B_6-catalysed reactions. *Journal of the American Chemical Society*, **76**, 648–652.

WIDER READING (BOOKS AND REVIEWS)

Dolphin, D., Poulson, R. and Avamic, O. (eds) (1987) *Pyridoxal Phosphate* Part A and Part B, Wiley-Interscience, New York.
Dolphin, D., Poulson, R. and Avamic, O. (eds) (1987) *Pyridine Nucleotide Coenzymes*, Part A and Part B, Wiley-Interscience, New York.
Jenks, W.P. (1987) *Catalysis in Chemistry and Enzymology*, Dover, New York.
Luckner, M. (1990) *Secondary Metabolism in Microorganisms, Plants, and Animals*, 3rd edn, Springer, Berlin.

CHAPTER 5

Pyrimidines

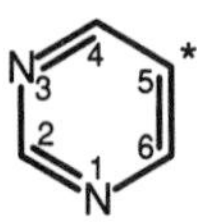

5.1 DISCOVERY AND OCCURRENCE

Although today the term 'pyrimidine' immediately brings to mind the nucleic acid bases uracil, cytosine and thymine, the first naturally occurring pyrimidine to be isolated was in fact a plant secondary product, vicine (1). This pyrimidine glucoside was isolated from seeds of *Vicia sativa* by Ritthausen in 1870 and later from several other species of *Vicia*, but it was not until more than eighty years later that its structure was elucidated (Bendich and Clements, 1953). Vicine is nearly always accompanied in Nature by a closely related pyrimidine glucoside, also discovered by Ritthausen (1881) and named by him convicine. The structure of convicine (2) was elucidated, much later again, by Bien *et al.* (1968).

(1)

(2)

The first pyrimidine constituent of nucleic acid to be discovered was thymine, isolated by Kossel and Neumann (1893) from hydrolysates of nucleic acid from calf thymus and beef spleen. During the following year, they isolated, from the same sources, a basic substance which they

*Until the late 1960s, the pyrimidine ring was conventionally numbered counterclockwise. This can cause confusion in searching older publications, e.g. uracil (2,4-dihydroxypyrimidine) was formerly 2,6-dihydroxypyrimidine.

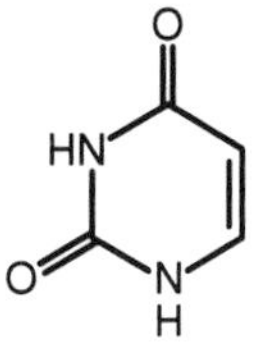

(3) 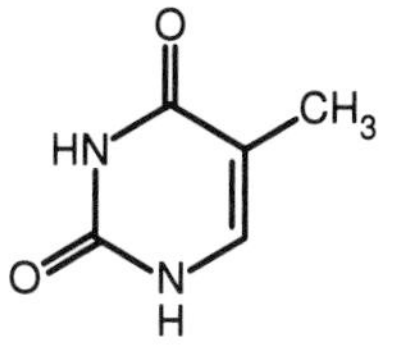 (4) 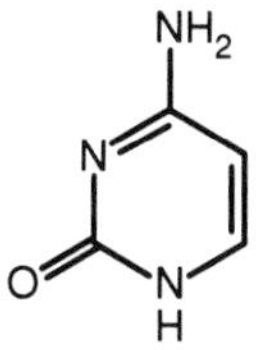 (5)

called 'cytosine' (Kossel and Neumann, 1894). The structure of thymine (3) was elucidated by Steudel (1900) and that of cytosine (4) by Kossel and Steudal (1903). The third pyrimidine base, uracil, was obtained from yeast nucleic acid by Ascoli (1900) and its structure (5) confirmed by unequivocal synthesis (Fischer and Roeder, 1901).

In addition to these three nucleic acid bases, smaller amounts of pyrimidines are found in most nucleic acids. These minor bases include 5-methylcytosine, found in fish, mammalian, and wheat-germ DNA (Hotchkiss, 1948; Wyatt, 1950, 1951). In T-even bacteriophages, 5-hydroxymethylcytosine replaces cytosine (Wyatt and Cohen, 1952). Some of the more commonly found minor bases of RNA are listed in Table 5.1. These generally comprise less than 5% of the total base content of RNA.

Uracil, cytosine and thymine also occur as free nucleosides and nucleotides in all living tissues (Table 5.2) but the free bases themselves, despite their involvement as intermediates in metabolism, seldom accumulate to any significant extent. Early reports describing free pyrimidine bases in tissues were often attributable to inappropriate extraction procedures, such as prolonged exposure to hot acid, which hydrolysed nucleic acids, nucleotides and nucleosides. In some cases, the extractants used (e.g. aqueous ethanol) failed to inactivate enzymes causing significant autolysis during the extraction process. The latter problem was particularly prevalent in studies of plant tissues in which widely distributed, non-specific phosphatases were subsequently shown not be inactivated by ethanol, even boiling ethanol (Bieleski, 1964). Still worse,

Table 5.1 Some of the more commonly occurring minor pyrimidine bases in RNA

4-acetylcytosine	5-methylaminomethyl-2-thiouracil
5-carboxymethyluracil	3-methyluracil
5,6-dihydrouracil	5-methyluracil
5-hydroxyuracil	5-methylamino-2-thiouracil
5-hydroxymethyluracil	5-methyl-2-thiouracil
5-hydroxymethylcytosine	5-methylcytosine
3-methylcytosine	2-thiocytosine
N^4-methylcytosine	4-thiouracil
4-methylcytosine	uracil-5-oxyacetic acid
5-methylcytosine	3-(3-amino-3-carboxypropyl)uracil

Table 5.2 The major pyrimidine ribonucleosides and ribonucleotides

Base	Ribonucleoside[2]	5'–Ribonucleotide [1,2,3]
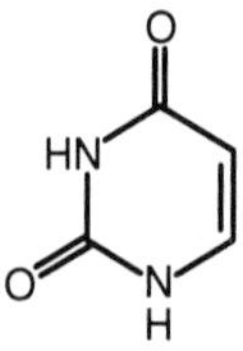 Uracil	Uridine 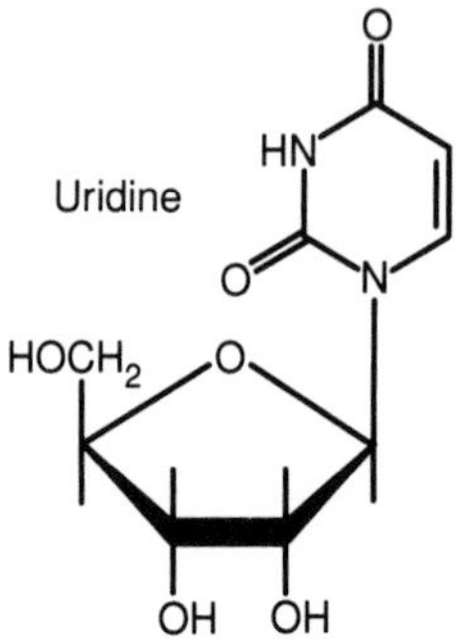	UMP 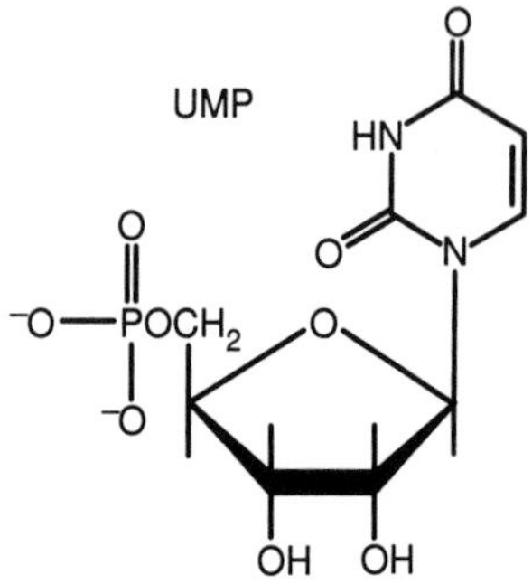
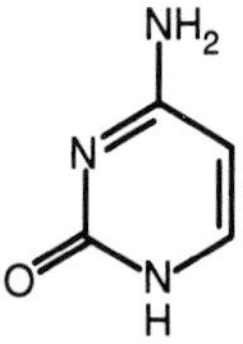 Cytosine	Cytidine 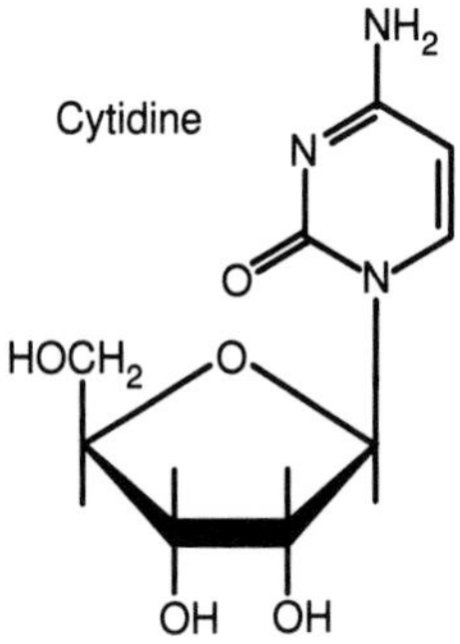	CMP 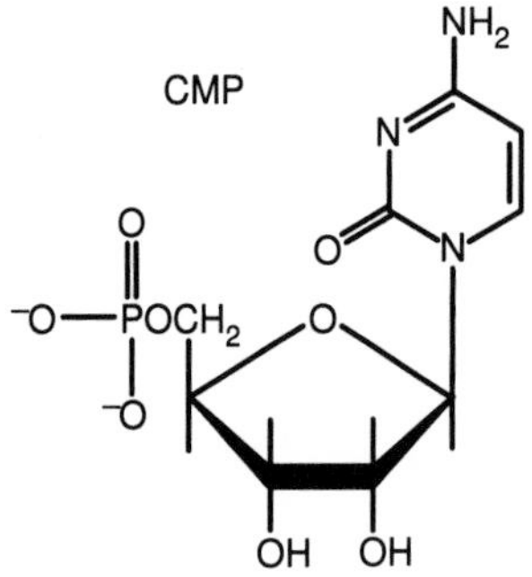
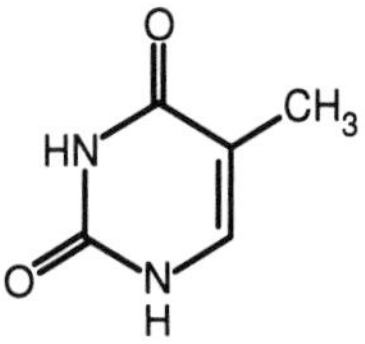 Thymine	Thymidine 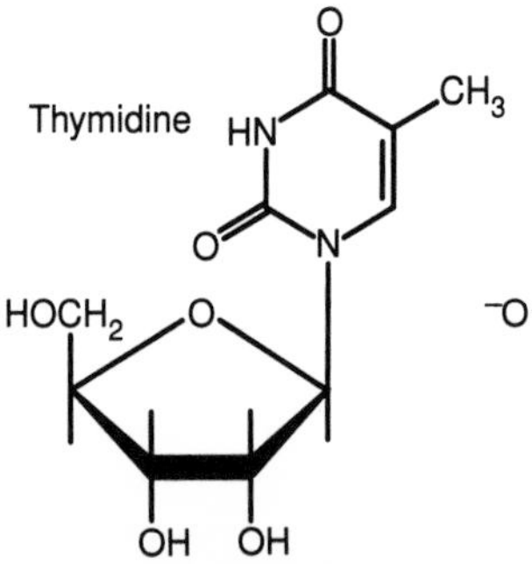	TMP 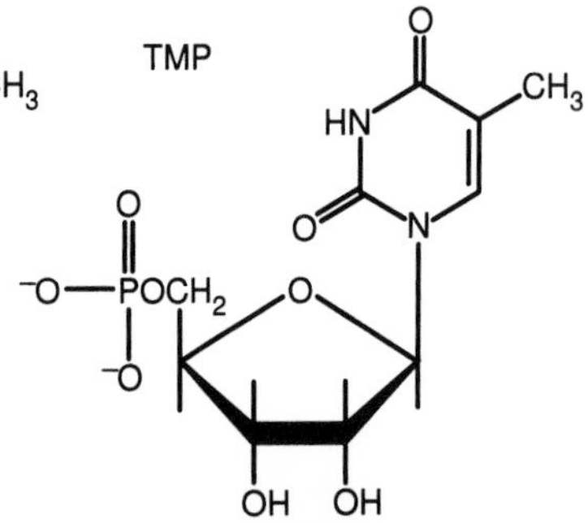

[1] Strictly speaking the terms 'nucleoside' and 'nucleotide refer to the compounds derived hydrolytically from nucleic acids but by common usage have come to have the wider connotation of all *N*–heterocyclic ribosides and ribotides of related general structure (e.g. nicotinamide mononucleotide, *q.v.*).

[2] The nucleotides and nucleosides derived from DNA are the 2'-deoxy derivatives).

[3] UMP, CMP and TMP are uridine 5'–monophosphate, cytidine 5'–monophosphate, and thymidine 5'–monophosphate, respectively.

such phosphatases can further complicate analytical data by catalysing transphosphorylation. These and related problems in extracting plant tissues have been discussed more fully by Brown (1991).

In addition to their combination with ribose or 2-deoxyribose, pyrimidines are found within the structure of a number of naturally occurring compounds, ranging from the vitamin thiamine (6) to antibiotics such as Blastodicin S (7). The pyrimidine ring is also found in combination with the amino acid alanine in the plant pyrimidine amino acids, willardiine (8), isowillardiine (9) and lathyrine (10).

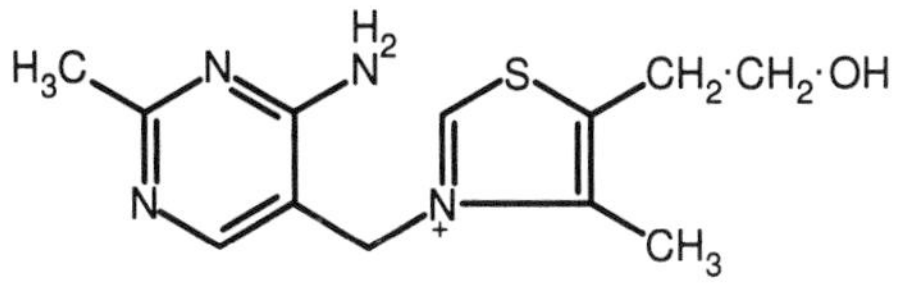

H_3C N NH_2 S $CH_2 \cdot CH_2 \cdot OH$ N N^+ CH_3

(6)

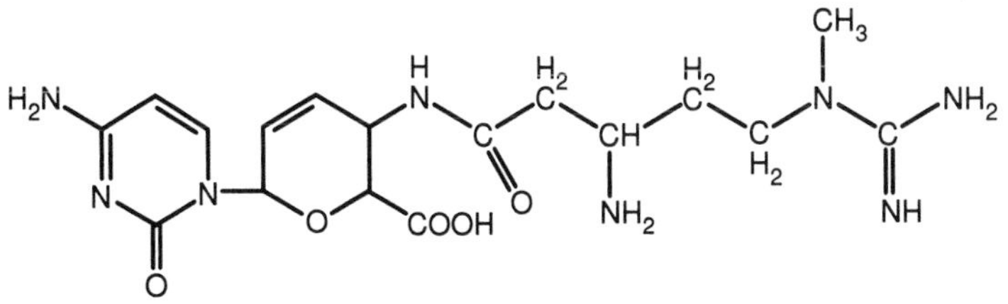

H_2N N N O O COOH H N C O H_2 C CH NH_2 H_2 C C H_2 CH_3 N C NH NH_2

(7)

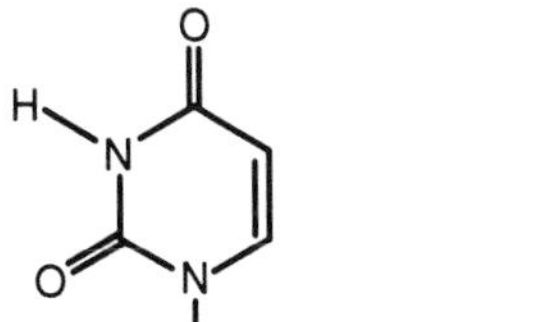

O H N O N $CH_2CH(NH_2)CO_2H$

(8)

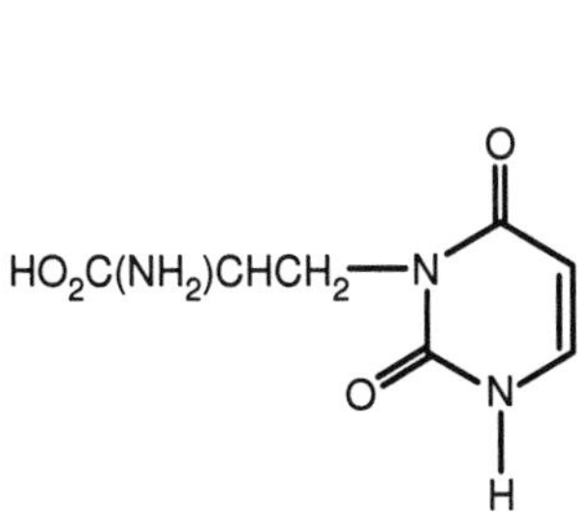

$HO_2C(NH_2)CHCH_2$—N O O N H

(9)

NH_2 CH_2CHCO_2H N H_2N N

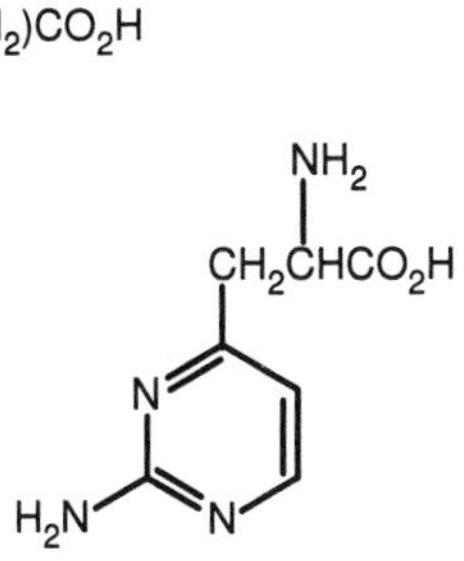

(10)

5.2 CHEMICAL PROPERTIES OF BIOCHEMICAL INTEREST

All known naturally occurring pyrimidine derivatives are crystalline solids. They are mostly amino- and/or hydroxy- derivatives, exhibiting aromaticity and tautomerism. Under physiological conditions the hydroxypyrimidines, such as uracil, exist predominantly in the oxo (lactam) form (Fig. 5.1) whereas the corresponding amino compounds do not exist to any appreciable extent in the imino form. Despite the predominance of the oxo-forms, because it is simpler and unambiguous, pyrimidine nomenclature commonly uses the hydroxy designations, e.g. referring to uracil as 2,4-dihydroxypyrimidine.

Like benzene, pyrimidines exhibit selective absorption in the UV-region of the spectrum. The resulting characteristic spectral properties are particularly useful in biochemistry and chemistry, not only facilitating identification but enabling rapid quantitative determinations to be made spectrophotometrically. It needs to be emphasized, however, that pyrimidines can exist in solution in a number of ionic forms depending on the pH of the solution and on their own individual pK_a values. Spectra recorded in an aqueous solution of arbitrary pH may well represent the summation of the spectra of a number of coexisting ionic forms. It is essential therefore not only to define the pH of the solution but to choose one which is at least two units above or below the pK_a values (Table 5.3) for the pyrimidine concerned. This can cause problems in automatic monitoring of the absorption spectra of chromatographic column eluates in which pH is dictated by required chromatographic conditions rather than spectrophotometric considerations.

There is now extensive literature concerning the infra-red, nmr and mass spectra of pyrimidines but important though these techniques are in identifying novel compounds it is beyond the scope of this book to go into detail. A good general account is to be found in Hurst (1980) or in the more specialized works on these individual analytical techniques, such as those by Waller and Dermer (1980); Schram (1989); and Knowles *et al.* (1976).

Although substituents can be oxidized readily, both chemically and

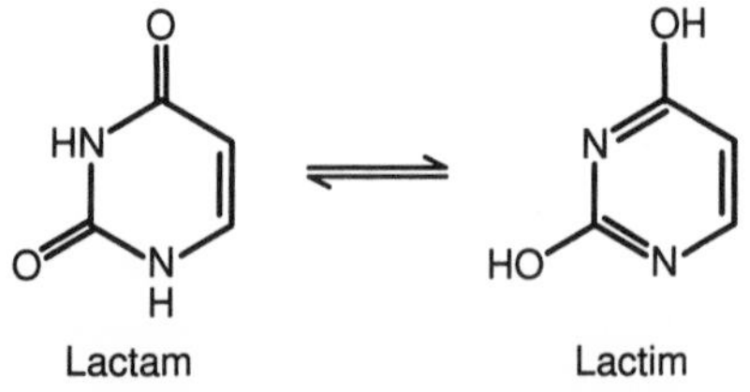

Fig. 5.1 Tautomerism of uracil. Under physiological conditions of pH (pH 6–7) the lactam form predominates.

Table 5.3 pK_a values of some common pyrimidine derivatives

	pK_a	
Compound	Base	Phosphate
Cytosine	4.6 12.2	
Cytidine	4.2 12.5	
CMP	4.8	6.6
CDP	4.6	6.4
CTP	4.8	6.6
Thymine	9.9 >13.0	
Deoxythymine	9.8	
dTMP	10.0	1.6 6.5
Uracil	9.5 >13.0	
Uridine	9.2 12.5	
UMP	9.5	6.4
UDP	9.4	6.5
UTP	9.6	6.6
5-Ribosyl-uracil	9.0	
	>13.0	
Orotic acid	2.4 9.4	

biochemically, the pyrimidine ring itself is resistant to oxidation. It can however be reduced either by catalytic hydrogenation using Pt, Pd or Rb catalysis, or biochemically by the NADPH-linked enzyme dihydrouracil dehydrogenase. In both cases, i.e. chemically or biochemically, the 5,6 double bond is saturated and results in ring-lability with facile hydrolytic fission occurring between *N*-3 and *C*-4. Biologically, this is the mechanism used for initiating the catabolism of pyrimidines (Fig. 5.2).

An aspect of pyrimidine oxidation of current biochemical interest relates to the intracellular generation of reactive oxygen species. This

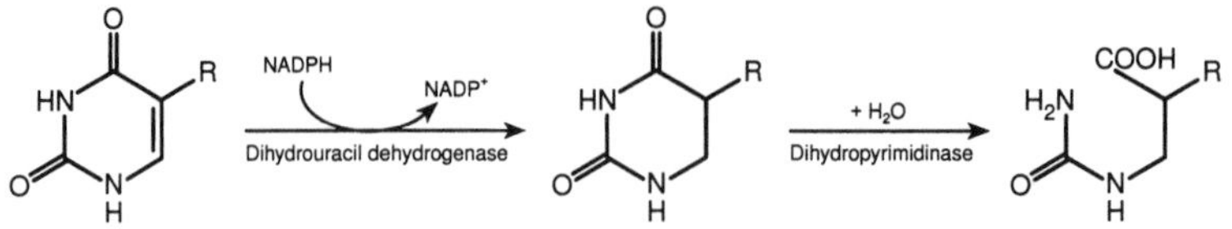

uracil (R = H)
thymine (R = CH_3)

NADPH → $NADP^+$ Dihydrouracil dehydrogenase

5,6-dihydrouracil (R = H)
5,6-dihydrothymine (R = CH_3)

+ H_2O Dihydropyrimidinase

N-carbamoyl β-alanine (R = H)
N-carbamoyl 2-methyl-3-aminopropanoic acid (R = CH_3)

Fig. 5.2 Enzymic fission of the pyrimidine ring. Reduction of the 5,6-double bond by NADPH, catalysed by dihydrouracil dehydrogenase, produces the labile 5,6-dihydro-derivative. This ring opens as shown. The same enzyme catalyses a parallel process with thymine.

can result in oxidation of thymine residues within DNA molecules and there is current speculation that the process, which results in significant urinary excretion of thymine glycol (10a), is a key process in both ageing and carcinogenesis.

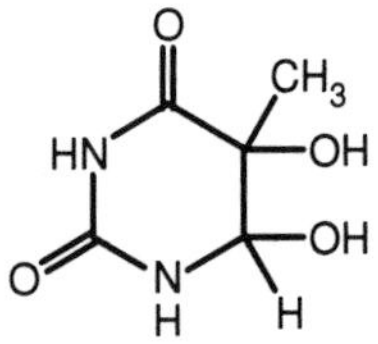

(10a)

The reaction of pyrimidines with hydroxylamine and with hydrazine (Fig. 5.3) are also of biological importance. Hydroxylamine is mutagenic, a property which is attributed to its action in hydroxylating the cytosine residues of DNA to form N^4-hydroxycytosine residues. It reacts differently with uracil, transforming the pyrimidine ring into that of an isoxazole. Hydrazine reacts with pyrimidines, causing a ring-contraction, to yield pyrazoles (Fig. 5.3). DNA in 2M NaCl is cleaved at cytosine residues by hydrazinolysis in this way; the thymine residues of DNA are however unaffected.

A number of other addition reactions can take place at the 5,6-double bond of the pyrimidine ring. Some of these have biochemical significance, especially in connexion with DNA structure and function. For example, in aqueous solutions of uracil or cytosine that have been exposed to irradiation with UV-light, the 5,6-double bond is hydrated but the reaction is readily reversed by warming under acidic conditions (Fig. 5.4). In frozen solutions of uracil or thymine subjected to UV-irradiation, dimerization occurs, forming cyclobutanes (Fig. 5.5). This type of dimerization can occur in DNA molecules *in situ* between adjacent thymine residues on the same strand (Fig. 5.6). The dimer blocks the action of DNA polymerase and so prevents replication of the affected

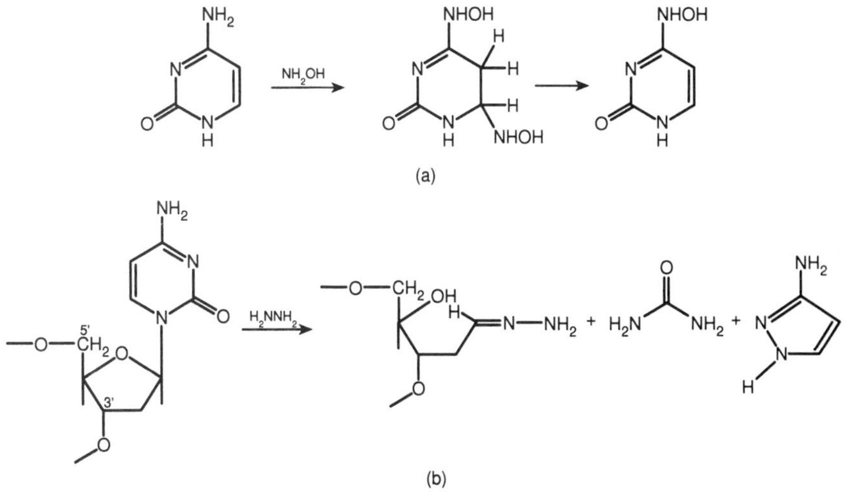

(a)

(b)

Fig. 5.3 (a) Hydroxylamine adds across the 5,6-double bond of cytosine yielding an adduct which, in turn, yields N^4-hydroxycytosine. This is the basis of the mutagenic effect of hydroxylamine. Cytosine residues in DNA are hydroxylated, as shown, and can no longer base-pair with guanine. (b) Hydrazine splits pyrimidine nucleosides, and results in ring-contraction of the pyrimidine base, yielding a pyrazole derivative.

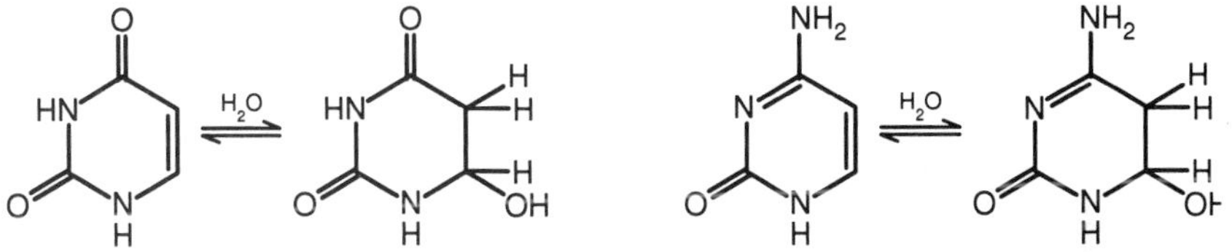

Fig. 5.4 Ultraviolet-light catalysed hydration of the 5,6-double bond of (a) uracil, and (b) cytosine. Warming under acidic conditions readily reverses the process.

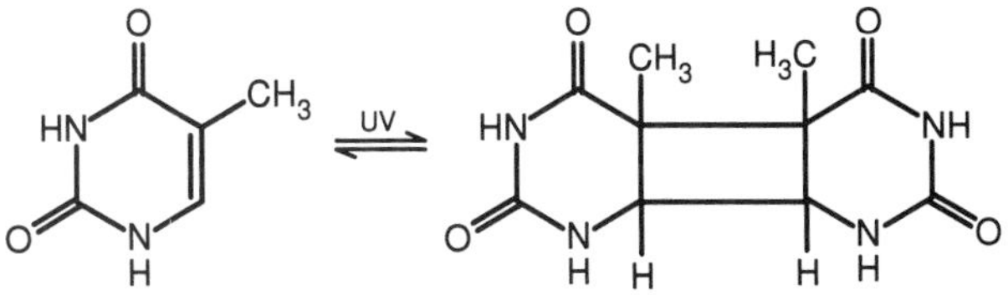

Fig. 5.5 Ultraviolet-light catalysed dimerization of thymine in frozen aqueous solution.

strand. This accounts, at least in part, for the damaging effect of UV-light on the genetic mechanism. More recently, UV-irradiation of DNA has been shown to result in mutagenesis by another mechanism involving two adjacent thymine residues on the same DNA strand. This

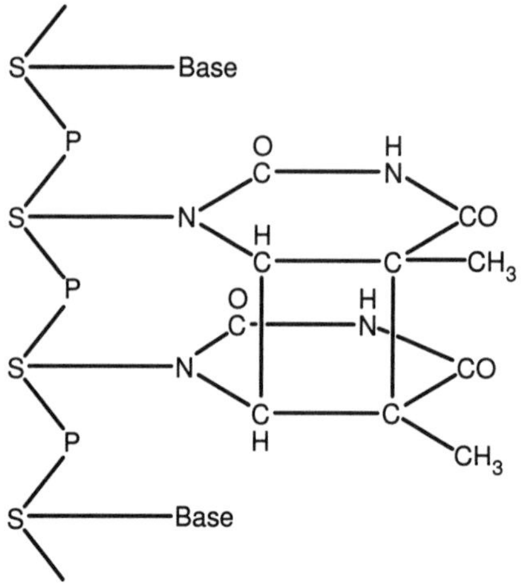

Fig. 5.6 Cyclobutane thymine dimer formed by adjacent thymine residues following irradiation of DNA with ultraviolet light. –S–P–S– represents the sugar-phosphate backbone of the affected strand.

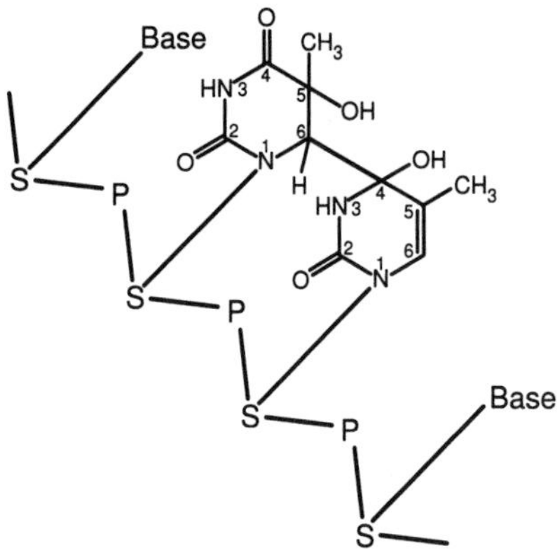

Fig. 5.7 The 6,4 covalent linkage formed by two adjacent thymine residues in DNA following irradiation with ultraviolet light. –S–P–S represents the sugar-phosphate backbone of the affected nucleic acid strand.

involves formation of a covalent bond between the 6-position of one thymine and the 4-position of its neighbour (Fig. 5.7).

5.3 BIOSYNTHESIS AND INTERCONVERSION

The major route for pyrimidine biosynthesis in animals, plants and microorganisms is the orotate pathway (Fig. 5.8). This sequence of six reactions utilizes, as elementary precursors, aspartate, CO_2 and the

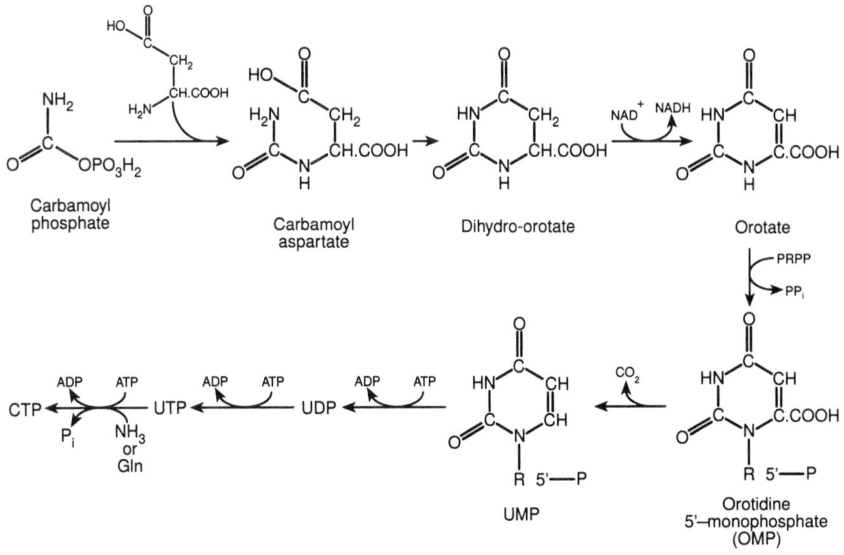

Fig. 5.8 The orotate pathway of pyrimidine biosynthesis. Gln represents glutamine.

amide-*N* of glutamine; ATP is also needed for the initial step, the generation of carbamoyl phosphate. The enzyme catalysing formation of carbamoyl phosphate for pyrimidine biosynthesis in animal tissues is a cytoplasmic carbamoyl phosphate synthetase (CPSase II) differing in properties and function from the mitochondrial carbamoyl phosphate synthetase (CPSase I) which has a requirement for *N*-acetylglutamate and is primarily concerned with the provision of carbamoyl phosphate for the urea cycle. In the plants and bacteria that have been investigated (e.g. O'Neil and Naylor, 1969, 1976, 1978; Ong and Jackson, 1972; Makoff and Radford, 1978) the carbamoyl phosphate requirements for both arginine (urea cycle) and pyrimidine synthesis are met by a single synthetase. The fungi, exemplified by *Saccharomyces, Aspergillus,* and *Neurospora* species, appear to be different again in this respect. They have two separate synthetases for the production of carbamoyl phosphate and the product of either is available for both arginine and pyrimidine synthesis (Makoff and Radford, 1978).

The carbamoyl phosphate synthetase (CPSase II) involved in pyrimidine biosynthesis in animal tissues is part of a multienzyme complex (*pyr* 1–3) whose other enzymic components are aspartate transcarbamoylase and dihydroorotase. This enzymic complex consists of three identical polypeptide chains, each of M_r about 230 000. The active site of the carbamoyl phosphate synthetase is unstable in aqueous buffers but is stabilized by cryoprotectants such as glycerol and ethylene glycol. It requires a stoichiometry of 2ATP for each molecule of carbamoyl phosphate synthesized, which has led to the suggestion that it catalyses a two-step process involving an ATP-dependent synthesis of carbamate from HCO_3^- and NH_2^-, followed by a kinase step resulting in the phosphorylation of enzyme-bound carbamate. Once dihydroorotate has been formed by the *pyr* 1–3 complex, it has to be dehydrogenated to orotic acid. Curiously, the dehydrogenase responsible is located in the mitochondria whereas all the other enzymes of the pathway are cytosolic.

In the animal pyrimidine-synthesizing system, the next two enzymes in the biosynthetic sequence (orotate phosphoribosyl transferase and OMP decarboxylase) are also part of a multienzyme complex (*pyr* 5,6) which is the product of a single gene. These enzymic activities catalyse, respectively, the ribotylation of orotate to OMP, and its subsequent decarboxylation to UMP. It is from the latter that the other common pyrimidine derivatives arise (Fig. 5.8).

2′-DeoxyCTP (dCTP), required for DNA synthesis, is produced from its corresponding ribonucleoside 5′-diphosphate by a reduction catalysed by ribonucleotide reductase and using thioredoxin and glutathione as reductants. This is followed by a further phosphorylation with ATP and involving the enzyme nucleoside diphosphate kinase, to yield the triphosphate. 2′-Deoxythymidine triphosphate (dTTP) is produced from dUTP, which arises in a parallel way to dCTP (Fig. 5.9). Two additional enzymic steps are however required. First, dUTP is hydrolysed to the monophosphate (dUMP) and then this is methylated at C-5 in a process using N^5,N^{10}-methylene tetrahydrofolate as the source of the methyl group. These mechanisms for synthesis of the pyrimidine deoxyribonucleotides are outlined in Fig. 5.9.

The regulation of pyrimidine biosynthesis in animal and microbial systems has been the subject of intensive research over a period of some years. Bacteria largely regulate their pyrimidine production through feedback control of the synthesis of carbamoyl aspartate. The key bacterial enzyme aspartate transcarbamoylase (ATCase) is probably the most intensively researched allosteric enzyme. It is inhibited by the end-product CTP and activated by ATP, increasing concentrations of which signal energy-rich cell conditions under which pyrimidines will be in demand for RNA and DNA synthesis. The enzyme exhibits quaternary structure and has six catalytic and six regulatory subunits. Each of the latter has one site capable of binding **either** CTP **or** ATP; regulation is thus a function of the CTP:ATP ratio in the cell.

In contrast, in the mammalian system, ATCase is not under allosteric control and it does not catalyse the first committed step in UMP biosynthesis, rather it is the CPSase that fills this function. The main regulated enzyme in the pyrimidine synthetic machinery of mammalian tissues is CPSase II, which is inhibited by UMP, UDP, UTP, CTP and UDP-glucose. The ATCase of higher plants appears to be of similar size to the catalytic subunit of the ATCase of *E. coli* and like the bacterial

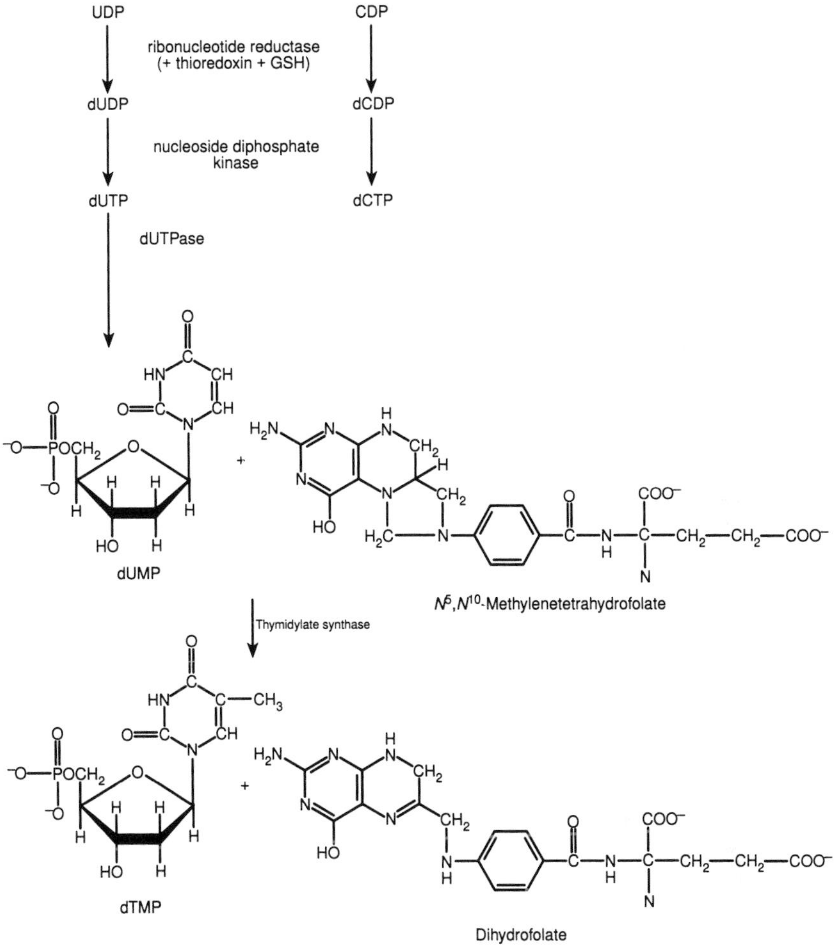

UDP → dUDP (ribonucleotide reductase (+ thioredoxin + GSH)) CDP → dCDP

dUDP → dUTP (nucleoside diphosphate kinase) dCDP → dCTP

dUTP → dUMP (dUTPase)

dUMP + N^5,N^{10}-Methylenetetrahydrofolate → dTMP + Dihydrofolate (Thymidylate synthase)

Fig. 5.9 Synthesis of pyrimidine 2′-deoxyribonucleotides. The nucleoside diphosphates are reduced by ribonucleotide reductase using thioredoxin and glutathione (GSH). The 2′-deoxyribonucleoside diphosphate is then further phosphorylated, ATP and a kinase being required. The resulting deoxyribonucleotides are used for DNA synthesis. To form 2′-deoxythymidine monophosphate, dUTP is hydrolysed to dUMP and this is methylated at C-5 by the enzyme thymidylate synthase using N^5,N^{10}-methylenetetrahydrofolate as the methyl-donor.

enzyme, it is allosteric. It is however primarily feedback-inhibited not by CTP but by UMP. It has been concluded by Wasternack (1982) that the pyrimidine biosynthesis of higher plants is controlled effectively through both of these allosteric enzymes (i.e. CPSase and ATCase). It is of interest in this context that both plant enzymes are feedback inhibited by UMP.

There are two other biochemical mechanisms known to synthesize the pyrimidine ring system but both of these are special cases rather than general biosynthetic routes. The first of these is the production of the pyrimidine moiety of thiamine (6). Supplying cultures of yeast with simple ^{14}C-labelled precursors, Tomlinson *et al.* (1966) and David and Estramareix (1977) found that there were significant inconsistencies between the labelling pattern in the thiamine produced and that expected if the pyrimidine ring had arisen from the operation of the orotate pathway of pyrimidine biosynthesis as outlined in Fig. 5.8. Studies with mutant strains of *Salmonella typhimurium* by Newell and Tucker (1968) subsequently indicated that the pyrimidine moiety of thiamine arises from a branch point in purine biosynthesis (Fig. 6.4) from which the intermediate 5-aminoimidiazole ribotide is diverted, and ring-expanded (Fig. 5.10).

The second special mechanism leading to formation of a pyrimidine ring occurs during the biosynthesis of the water-soluble vitamins folic acid and riboflavin (Figs. 7.3 and 8.3). In the case of both of these compounds, the imidazole ring of the purine precursor is opened, by enzymic removal of C-8, to form a 5,6-diaminopyrimidine intermediate (Fig. 5.11).

In higher plants, a number of secondary products arise during pyrimidine metabolism, e.g. the pyrimidine amino acids willardiine (8), isowillardiine (9), and lathyrine (19), and the pyrimidine glucosides vicine (1) and convicine (2). The indications are that all of these derive biosynthetically from uracil, indeed the formation of willardiine and isowillardiine (Ashworth *et al.*, 1972; Ahmmad *et al.*, 1984) results from the direct transfer of an alanine residue from *O*-acetylserine to uracil (Fig. 5.12). This is an enzyme-catalysed process which requires pyridoxal phosphate as a coenzyme (Ahmmad *et al.*, 1984).

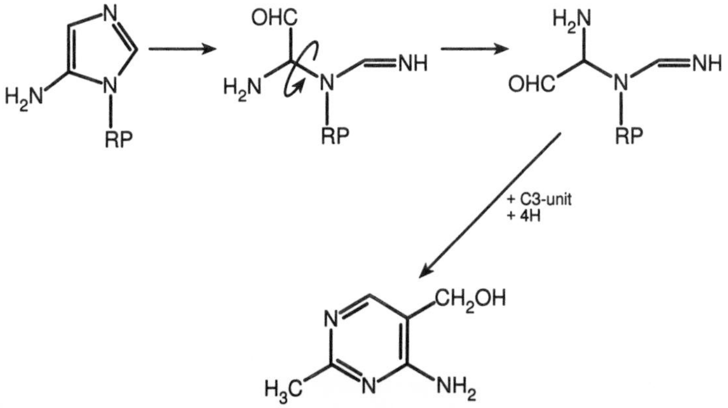

Fig. 5.10 Ring expansion of 5-aminoimidazole ribotide to form the pyrimidine moiety of thiamine. The nature of the three-carbon unit is unknown.

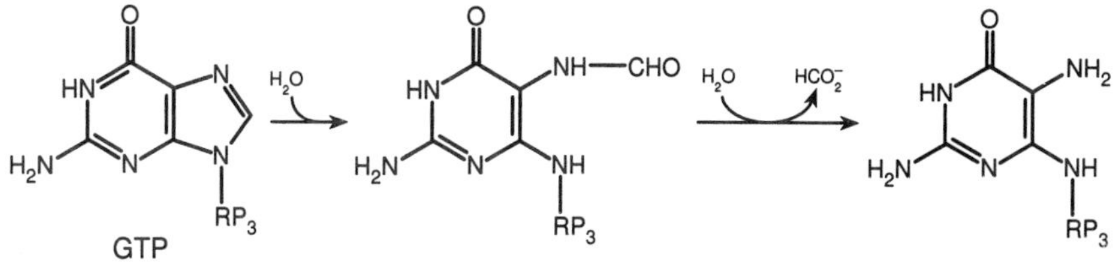

Fig. 5.11 During the biosynthesis of the vitamins riboflavin and folic acid, a diaminopyrimidine intermediate is formed by the tetrahydrofolic acid-dependent enzyme GTP-cyclohydrolase which takes C-8 out of the purine ring.

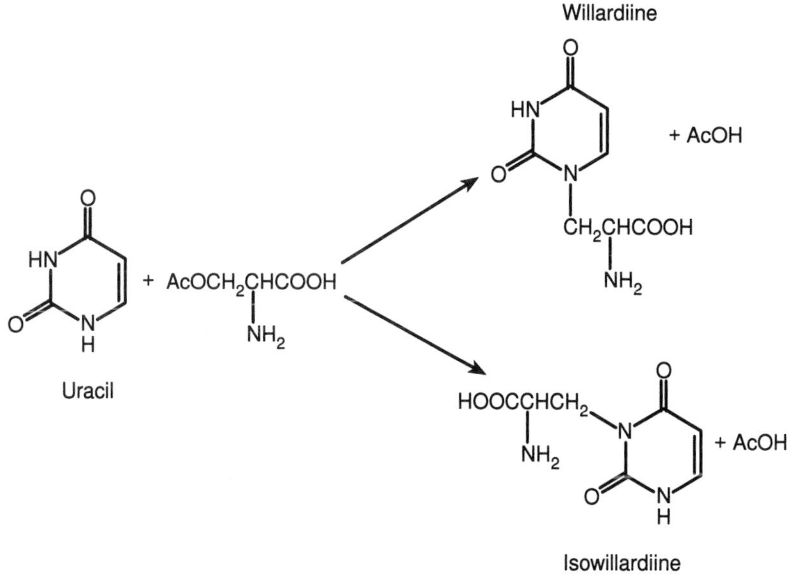

Fig. 5.12 Enzymic formation of willardiine and isowillardiine from uracil by direct transfer of an alanine residue from *O*-acetylserine.

On the basis of the observed incorporation of radioactivity from [^{14}C]homoarginine into lathyrine (10) by *L. tingitanus*, but without chemical degradation to locate the incorporated isotope, it had been postulated (Bell, 1963; Nowacki and Nowacka, 1963; Bell and Przbylska, 1965) that this pyrimidine amino acid is biosynthesized by the cyclization of 4-hydroxyhomoarginine (Fig. 5.13). It was, however, later shown by more detailed studies with ^{14}C-labelled orotic acid, uracil and serine, in which there was substantial incorporation of these precursors (Brown and Al-Baldawi, 1977) that the major biosynthetic route for lathyrine is *via* the orotate pathway of pyrimidine biosynthesis (Fig. 5.8) with the alanine moiety arising from serine. These latter studies were confirmed by chemical degradation of the biosynthesized

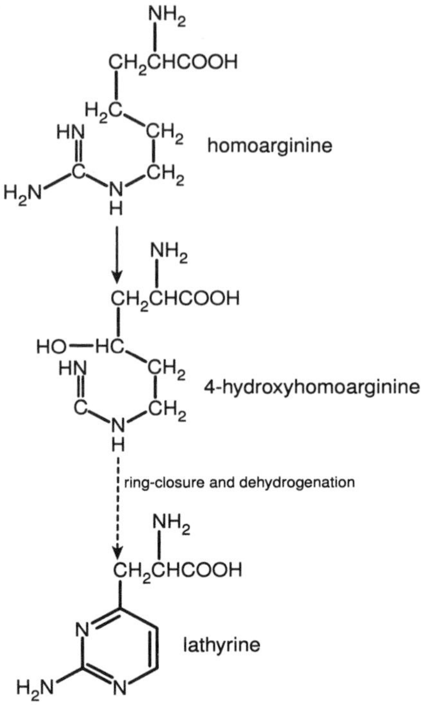

Fig. 5.13 Pathway originally postulated for the biosynthesis of lathyrine by the cyclization of 4-hydroxyhomoarginine.

lathyrine to locate the position within the molecule of the incorporated isotope from each precursor. Repeating the earlier experiments of Bell and his co-workers with [^{14}C]homoarginine, Brown and Al-Baldawi (1977) noted some minor incorporation of radioactivity into lathyrine. On a molar basis this was less than 4% of that obtained with ^{14}C-orotate of similar specific activity and they concluded that it results from the exogenously supplied [^{14}C]homoarginine feeding back through the first two reactions of pyrimidine catabolism (Fig. 5.14) by the mass action effect of the exogenously supplied hydroxyhomoarginine (Fig. 5.14). Surprisingly, over 20% of the orotate metabolized by *Lathyrus tingitanus* seedlings was converted into lathyrine (Al-Baldawi and Brown, 1983).

More recent studies by Brown and Mohamad (1990) have confirmed that the biosynthetic route for lathyrine involves a pyrimidine precursor arising *via* the orotate pathway. Uracil appears to be the link compound and is used to produce the immediate precursor of lathyrine which has been identified as 2-amino 4-carboxypyrimidine (Brown and Mohamad, 1990; Brown, 1996; Brown and Turan, 1996). In these studies, an enzymic preparation was obtained, from seedlings of *Lathyrus tingitanus*,

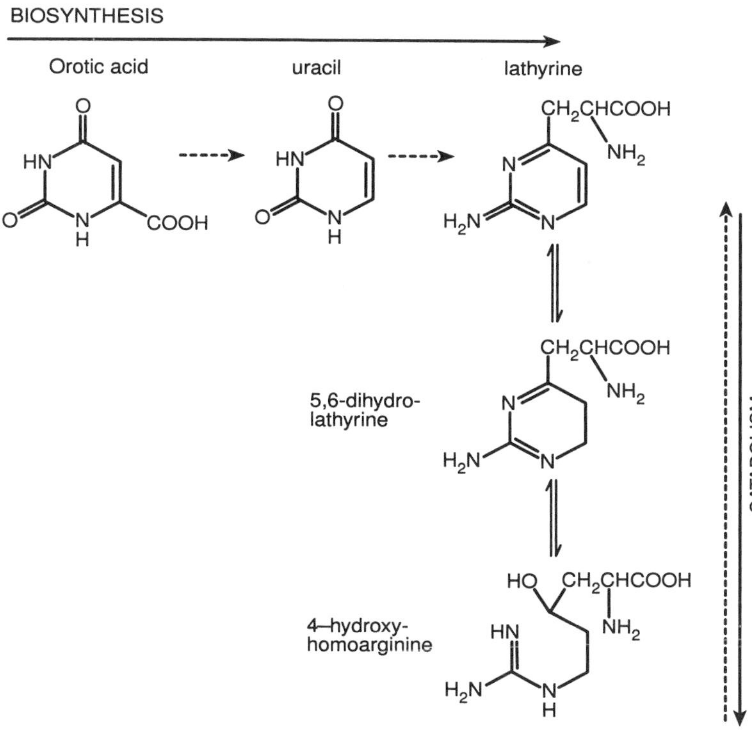

Fig. 5.14 Relationship demonstrated between lathyrine biosynthesis *via* the orotate pathway and incorporation of small amounts of radioactivity from [^{14}C]-4-hydroxyhomoarginine into lathyrine.

which catalyzes the simultaneous decarboxylation of this compound and the transfer to it of an alanine residue from serine (Fig. 5.15). The enzyme requires pyridoxal phosphate and unusually for a decarboxylation process, biotin. Unlike the enzyme catalysing the transfer of alanine to uracil to form willardiine and isowillardiine (Fig. 5.12) the lathyrine synthase specifically requires serine as the alanine donor and cannot use *O*-acetylserine.

Consideration of the biogenetic origin of 5-ribosyluracil (10b), also

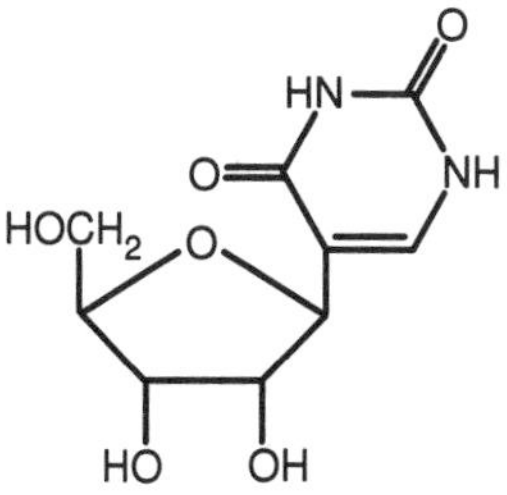

(10b)

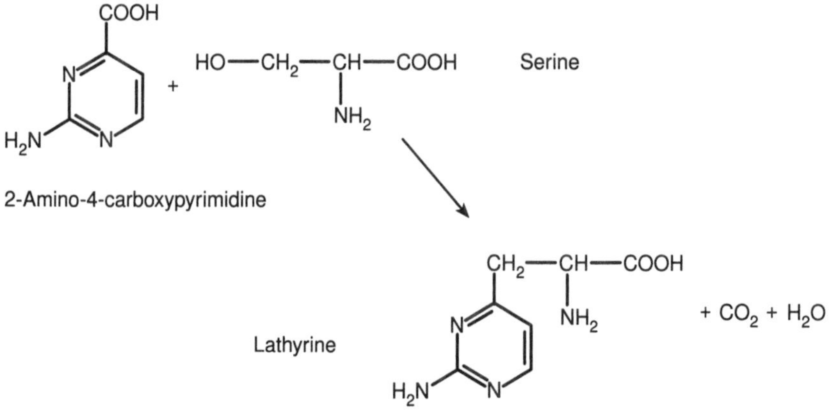

2-Amino-4-carboxypyrimidine + $HO—CH_2—CH(NH_2)—COOH$ Serine

→ Lathyrine + CO_2 + H_2O

Fig. 5.15 Simultaneous enzymic decarboxylation of 2-amino-4-carboxypyrimidine and transfer of an alanine residue from serine to form lathyrine.

known as pseudouridine (ψ-uridine), raises another unsolved problem of pyrimidine biochemistry. 5-ribosyluracil is an important constituent of various species of tRNA and rRNA. Structurally it is an isomer of uridine (Table 5.2) in which, unlike other naturally occurring pyrimidine nucleosides, the pentose is linked to a carbon atom (C-5) rather than to a nitrogen atom (N-1) in the pyrimidine ring. The molecule is usually thought of as a component of RNA where it is formed *in situ* by post-transcriptional modification of selected uridine residues. Enzymic preparations have been obtained that catalyse this process and the enzyme has been described as pseudouridine synthase. Samuelsson and Olsson (1990) have however presented evidence that there is more than one such enzyme. In studies of the post-transcriptional modification of a glycine tRNA by *Saccharomyces cerevisiae* they showed that uridine residues at three separate sites were converted to pseudouridine and that the enzymic activities concerned were chromatographically separable and specific to each site. Little is known however of the chemistry involved in this transglycosylation.

In addition to the occurrence of 5-ribosyluracil in RNA, it has also been found in a free state in some organisms. Cultures of some microorganisms even accumulate significant concentrations of the free nucleotide, e.g. *Agrobacterium tumefaciens, Saccharomyces cerevisiae,* and *Streptoverticillium ladakanus*. Al-Baldawi and Brown (1983) also described the accumulation of free 5-ribosyluracil within the tissues of bean seedlings (*Phaseolus vulgaris*). Their study, using ^{14}C-UTP, indicated that UTP is an intermediate in the formation of the accumulating compound but the results did not distinguish between its incorporation into RNA followed by catabolic release, and the possibility of direct synthesis of 5-ribosyluracil from the labelled precursor. The data did however exclude uracil, uridine and UMP as more immediate precursors. Working with *Streptoverticillium ladakanus* which accumulates 5-ribosyluracil in its culture medium, Uematsu and Suhadolnik (1973) found evidence for both direct synthesis and catabolic production from RNA.

Another minor nucleoside constituent of RNA that is of interest in a biosynthetic context is 3-(3-amino-3-carboxypropyl)uridine (10c). This was found to be a component of the phenylalanine-tRNA of *E. coli* and, structurally, is the riboside of the next higher homologue of the pyrimidine amino acid isowillardiine (9). The question that this structure poses is whether it is formed in a parallel reaction to the pyridoxal phosphate-dependent alanylation of uracil which forms willardiine and isowillardiine (Fig. 5.12). The work of Nishimura *et al.* (1974) indicates however that *S*-adenosylmethionine (10d) is the donor of the 3-amino-3-carboxypropyl group and that a uridine residue in tRNA is the acceptor, i.e. this is yet another example of the post-transcriptional modification of a pyrimidine.

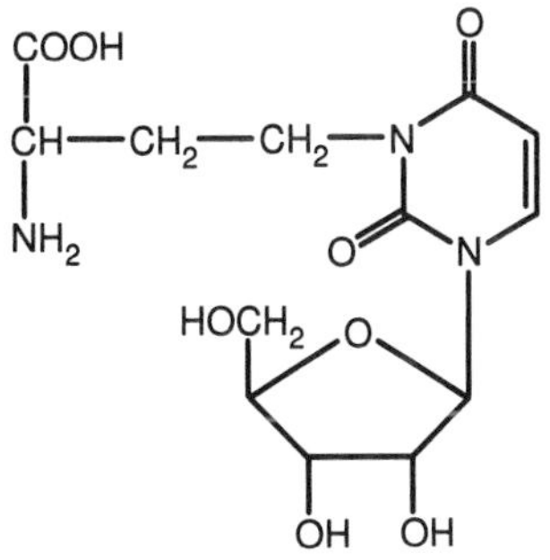

(10c) 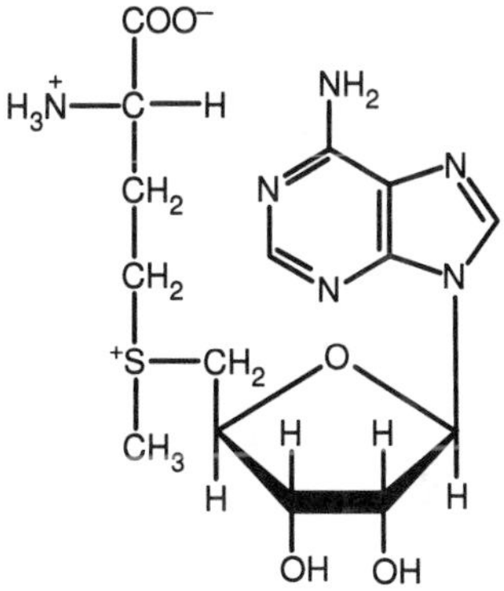 (10d)

As described above, the pyrimidine glucosides vicine (1) and convicine (2) are produced by various species of *Vicia*. They are the 5-*O*-glucosides, respectively of 2,6-amino-4,5-dihydroxypyrimidine and 6-amino-2,4,5-trihydroxypyrimidine and whereas their pyrimidine rings are known to be formed from orotate (Brown and Roberts, 1972), nothing is known of the enzymic systems involved in the necessary additional aminations and hydroxylations.

5.4 PYRIMIDINE CATABOLISM

In all the biological systems examined, pyrimidine catabolism is essentially similar to that shown in Fig. 5.16. Enzymic reduction of the 5,6-double bond, by NADPH-dependent dihydrouracil dehydrogenase, labi-

uracil ← cytosine thymine

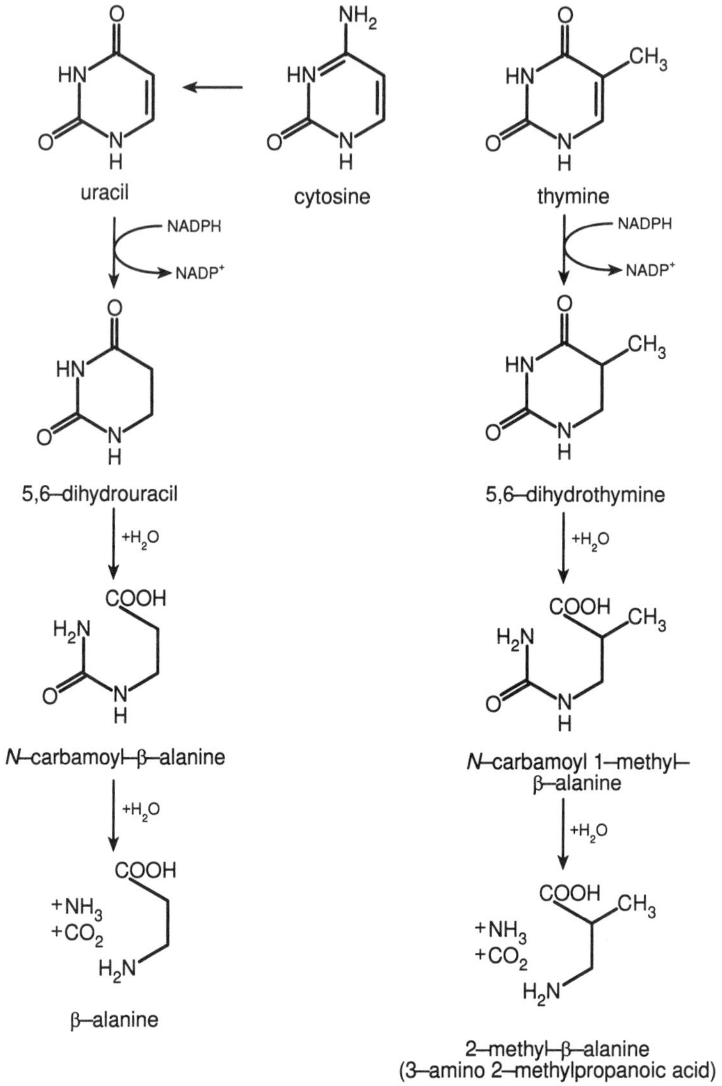

$NADPH \rightarrow NADP^+$

5,6–dihydrouracil 5,6–dihydrothymine

$+H_2O$

N–carbamoyl–β–alanine *N*–carbamoyl 1–methyl–β–alanine

$+H_2O$

$+NH_3$ $+CO_2$

β–alanine

2–methyl–β–alanine (3–amino 2–methylpropanoic acid)

Fig. 5.16 Catabolism of the common pyrimidine bases, uracil, cytosine and thymine.

lizes the bond between N-3 and C-4 and leads to hydrolytic ring fission, catalysed by 5,6-dihydropyrimidinase. The carbamoyl group of the product is removed by further enzymic hydrolysis resulting in liberation of CO_2 + NH_3. The end product in the case of uracil is β-alanine, with thymine it is 2-methyl β-alanine (3-amino-2-methylpropanoic acid). In animal tissues and microorganisms, cytosine is hydrolytically deaminated to uracil before degradation. Higher plants do not however

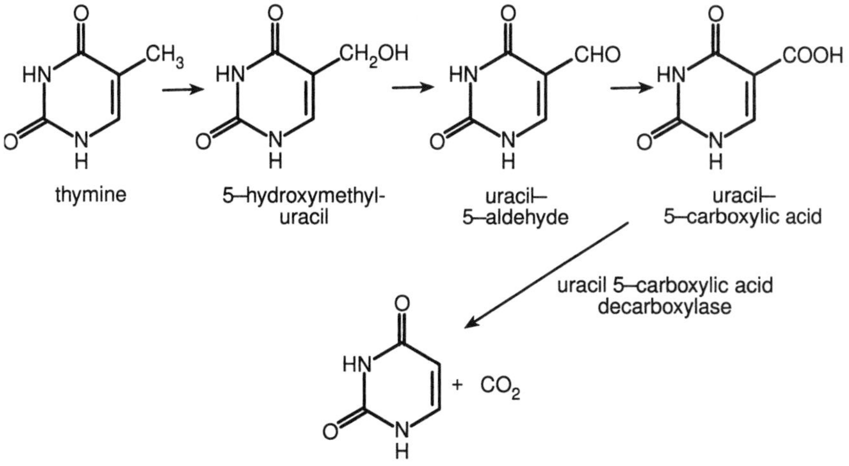

Fig. 5.17 Oxidative degradation of the 5-methyl group of thymine by *N. crassa*. The first three steps are catalysed by a mixed-function oxygenase, thymine 7-hydroxylase.

appear to possess the necessary cytosine deaminase for this to occur directly but the reported occurrence of cytidine deaminase implies that the necessary deamination is effected at the nucleoside level. In *N. crassa*, the 5-methyl group of thymine is eliminated following catabolism by an unusual series of oxidative reactions culminating in formation of 5-carboxyuracil. This process is catalysed by a mixed-function oxygenase, thymine 7-hydroxylase, and is followed by an enzymic decarboxylation to yield uracil; the sequence of reactions involved are shown in Fig. 5.17.

There were several well-documented reports some years ago, exemplified by that of Hayaisi and Kornberg (1952), that strains of *Corynebacterium* and *Mycobacterium* degrade pyrimidines oxidatively *via* barbituric acid according to the scheme shown in Fig. 5.18. As in the more usual catabolic pathway (Fig. 5.16) cytosine is first hydrolytically deaminated to uracil.

In some plant species, catabolism of pyrimidines leads to production of unusual ('non-protein') amino acids such as albizziine (2-amino-3-ureidopropanoic acid), 2,3-diaminopropanoic acid, and 4-hydroxyhomoarginine. The reason for this is that, although the pyrimidine catabolic pathway is essentially similar in most plants, animals and microorganisms, some plants produce pyrimidine secondary products that find their way into the catabolic machinery. The enzymes concerned show group specificity, i.e. they catalyse parallel reactions on a range of analogues of uracil and thymine (Fig. 5.19).

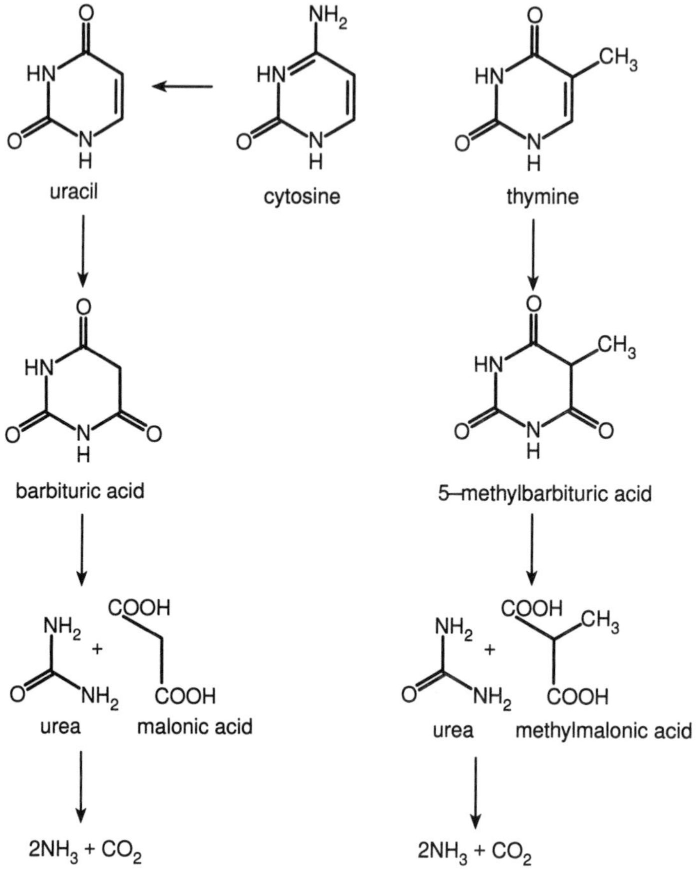

uracil, cytosine, thymine, barbituric acid, 5-methylbarbituric acid, urea, malonic acid, urea, methylmalonic acid, $2NH_3 + CO_2$, $2NH_3 + CO_2$

Fig. 5.18 Oxidative degradation of pyrimidines by species of *Corynebacterium* and *Mycobacterium*.

5.5 BIOCHEMICAL FUNCTIONS OF PYRIMIDINES

Pyrimidines function in biological systems primarily at the nucleotide level (Table 5.2). Like their purine counterparts, they are not only precursors of nucleic acids but function in their own right as allosteric regulators and coenzymes. During pyrimidine biosynthesis (Fig. 5.4) UMP concentration controls the activity of both carbamoyl phosphate synthetase and aspartate transcarbamoylase in plants. In animal tissues UDP, UTP and CTP are important regulators of CPSase II activity (Section C). In bacteria, ATCase is regulated allosterically by CTP.

The uridine nucleotide UDP-glucose plays a particularly prominent role in carbohydrate metabolism. Until the pioneering work of Luis Leloir and his colleagues in the 1950s, the interconversion of glucose

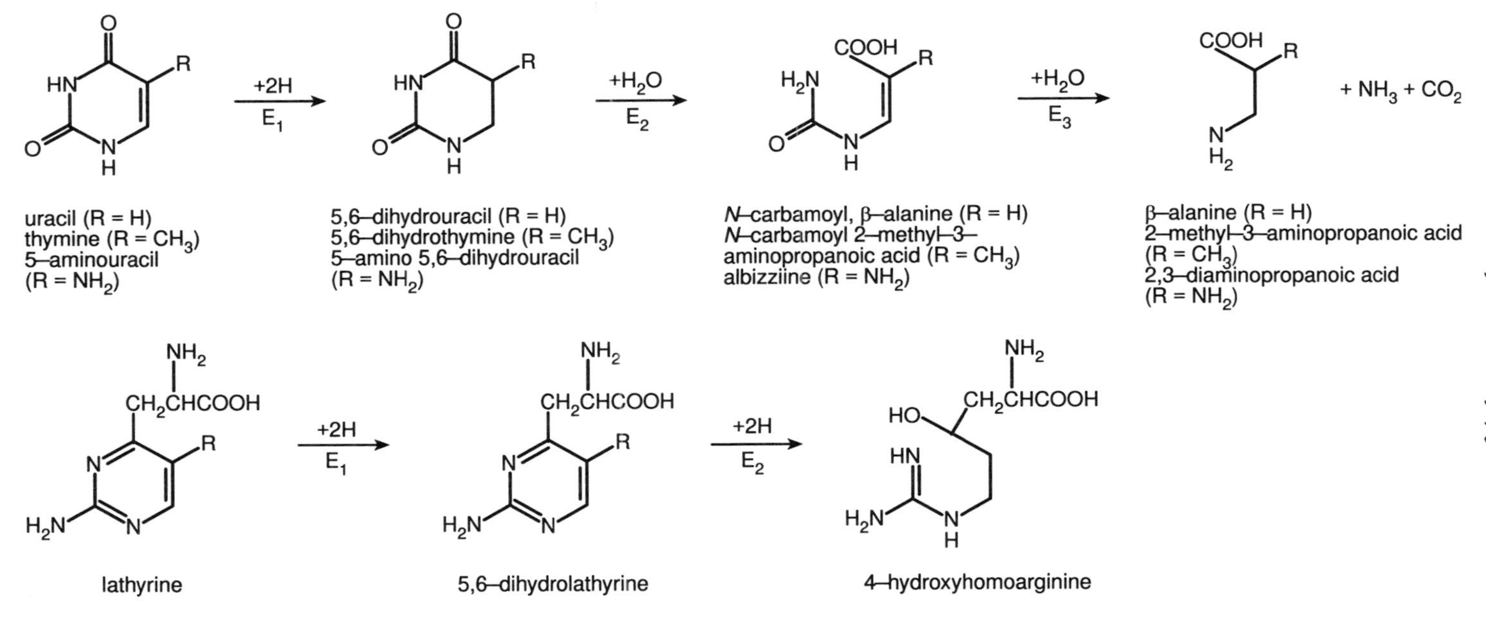

Fig. 5.19 Group specificity of the enzymes of the pyrimidine catabolic pathway results in the biogenesis of albizziine and 2,3-diaminopropanoic acid from 5-aminouracil, and of 4-hydroxyhomoarginine from lathyrine. E_1, dihydrouracil dehydrogenase; E_2, dihydropyrimidinase; E_3, β-ureidopropionase (Reprinted from *Phytochemistry*, Brown and Turan, vol. 40, pp. 763-771. Copyright 1995, with kind permission of Elsevier Science Ltd.)

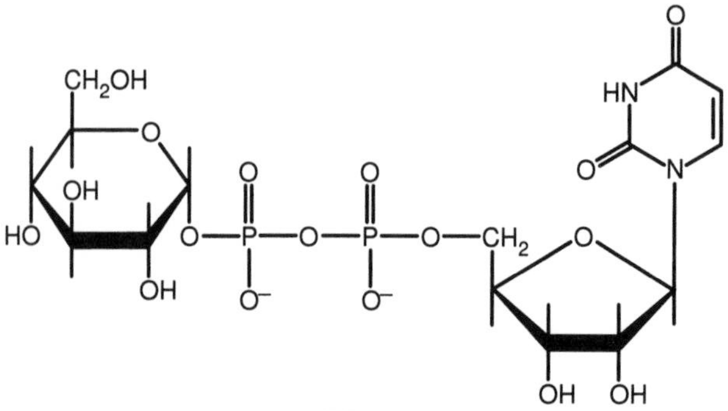

UDP-glucose

Fig. 5.20 Structure of the coenzyme UDP-glucose.

and galactose in liver and various other tissues was thought to be a single-step enzymic epimerization. Leloir showed that the process required UDP-glucose (Fig. 5.20) as a coenzyme and that several steps were involved. The process as it is now understood is outlined in Fig. 5.21. Further research into UDP-glucose and its analogues revealed that these compounds are important biological glycosylation agents, involved not only in the formation of a wide variety of naturally occurring glycosides but also as glycosyl donors in the biosynthesis of polysaccharides. Glycogen, for example, is formed by sequential enzymic glycosyl transfer from UDP-glucose to the non-reducing end of a primer chain, initially consisting of four or more α 1,4-linked glucose mono-

1. galactose + ATP $\xrightarrow{1}$ galactose 1–P + ADP
2. galactose 1–P + UDP–glucose $\xrightarrow{2}$ glucose 1–P + UDP–galactose
3. UDP–galactose $\xrightarrow[3]{NAD^+ \quad NADH}$ UDP–4–ketoglucose $\xrightarrow[3]{NADH \quad NAD^+}$ UDP–glucose
4. UDP–glucose + PP_i $\xrightarrow{4}$ UTP + glucose 1–P
5. glucose 1–P $\xrightarrow{5}$ glucose 6–P

Fig. 5.21 Enzymic conversion of galactose to glucose 6-phosphate. In the liver, the latter compound would normally be further metabolized but can also be easily hydrolysed by phosphatases to yield glucose. The process is used in reverse in mammary glands for the synthesis of galactose for lactose formation. The enzymes involved are: 1, galactokinase; 2, UDP-glucose:galactose 1-P uridyltransferase; 3, UDP-galactose 4-epimerase; 4, UDP-glucose pyrophosphorylase; 5, phosphoglucomutase.

mers. Similarly, in the biosynthesis of the glycoprotein components of connective tissue and in the glycosylation of newly synthesized proteins, UDP-glycosyl compounds play an important role. Although the full details of cellulose biosynthesis in higher plants are not established, here too it is almost certain that UDP-glucose is the immediate glucosyl donor. It is of interest that the sugars transported by UDP are almost always aldoses although the reasons for this are obscure. The exception that proves the rule is UDP-fructose, found in pea seedlings (Brown and Mangat, 1967) and in tubers of Jerusalem artichoke (Umemura, Nakamura and Funahashi, 1967). The biochemical significance of this compound is unknown and it does not appear to serve as a fructose donor in fructan biosynthesis, the key compound in which is the trisaccharide 1-kestose (see Pollock and Cairns, 1991).

A similar type of carrier-coenzyme relationship to that between UDP and aldoses exists between CDP and biologically important alcohols. This relationship is exemplified by CDP-choline and CDP-ethanolamine (Fig. 5.22), both of which are involved in glycerophospholipid metabolism in eukaryotes. With each of these cytidine nucleotides, the alcohol residue is transferred enzymically to diacylglycerol. In the case of CDP-choline, the product is phosphatidylcholine, and with CDP-ethanolamine phosphatidylethanolamine is formed (Fig. 5.23). Other CDP-alco-

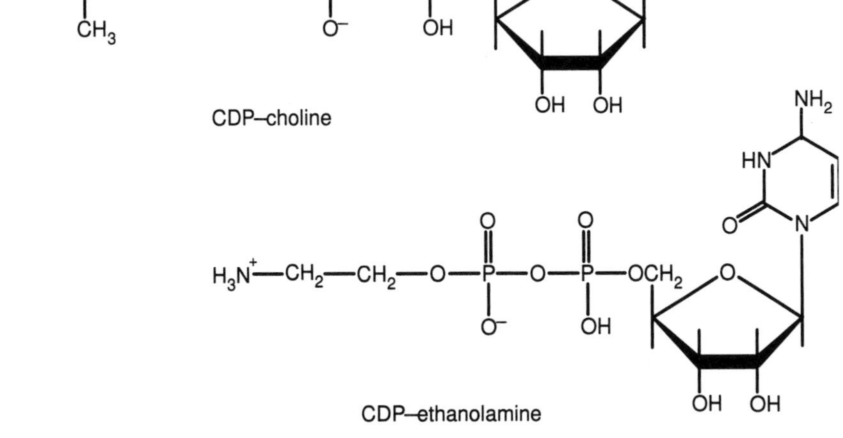

CDP–choline

CDP–ethanolamine

Fig. 5.22 Structure of CDP-choline and CDP-ethanolamine, both of which are involved in glycerophospholipid metabolism.

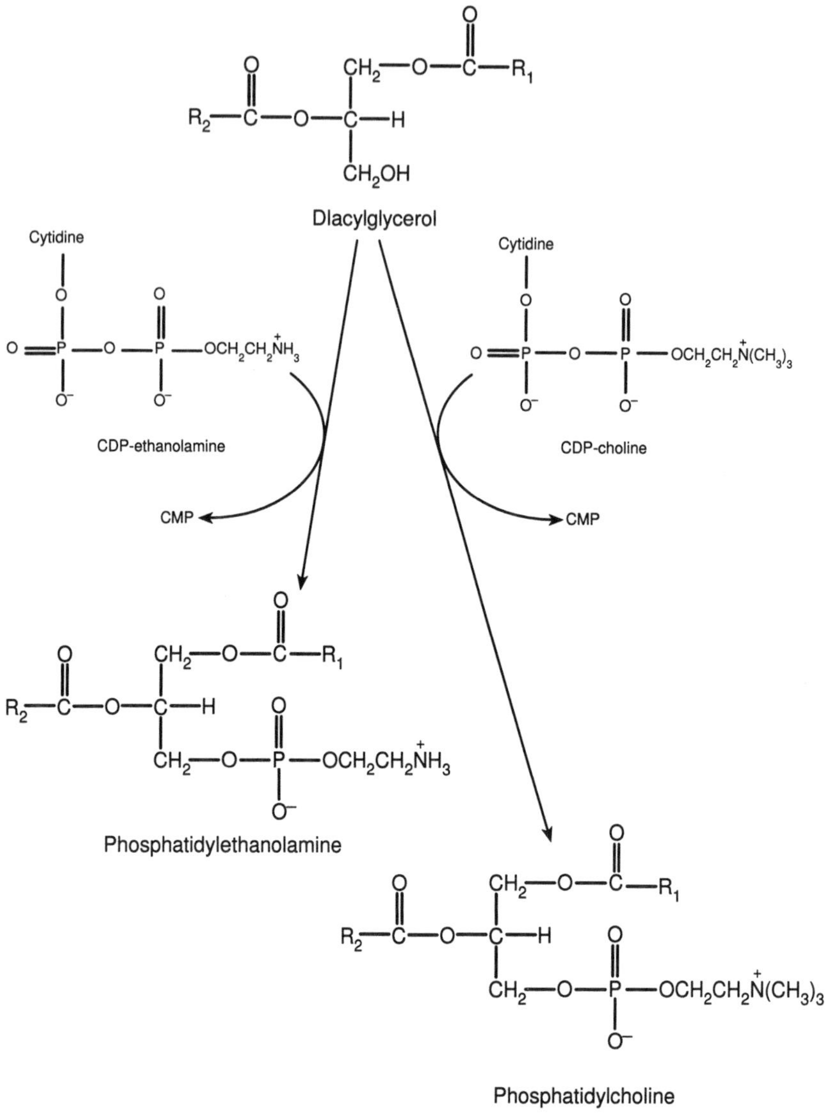

Diacylglycerol

Cytidine

CDP-ethanolamine

CMP

Phosphatidylethanolamine

Cytidine

CDP-choline

CMP

Phosphatidylcholine

Fig. 5.23 The role of CDP-ethanolamine and CDP-choline in glycerophospholipid synthesis. The formation of phosphatidylethanolamine is catalysed by CDP-ethanolamine:1,2-diacylglycerol phosphoethanolamine transferase. That of phosphatidylcholine is catalysed by CDP-choline 1,2-diacylglycerol phosphocholine transferase.

hols include CDP-ribitol and CDP-glycerol which are donors of their respective alcohol residues during the biosynthesis of teichoic acid required for bacterial cell wall formation.

Both groups of compounds, the UDP-sugars and the CDP-alcohols, can be formed (i) from the corresponding nucleoside 5′-triphosphate by enzyme catalysed reactions of the type:

$$\text{Nucleoside triphosphate} + \text{XP} \rightarrow \text{Nucleoside diphosphate-X} + \text{PP}_i$$

and (ii) from pre-existing nucleoside diphosphate-X compounds by transferase reactions:

$$\text{Nucleoside diphosphate-X} + \text{X'P} \rightarrow \text{Nucleoside diphosphate-X'} + \text{XP}$$

5.6 BIOLOGICAL AND PHARMACOLOGICAL ACTIVITY OF NATURALLY OCCURRING PYRIMIDINES

A number of naturally occurring pyrimdine derivatives exhibit biological activity. For example, the pyrimidine glucoside vicine (1) has been implicated as the causative dietary factor in the disease favism which is characterized by a potentially fatal haemolytic anaemia in genetically predisposed individuals with an hereditary deficiency of erythrocyte glucose 6-phosphate dehydrogenase activity. With susceptible individuals, bouts of the disease commonly follow ingestion of broad beans (*Vicia faba*); in particularly sensitive cases, even inhaling bean pollen has been reported to induce an attack. The disease has been known since antiquity and what is believed to be one of the earliest recorded cases is that of the Greek philosopher Pythagoras in the 6th century BC (London, 1961). Because of its genetic link, favism occurs primarily in specific geographical areas, e.g. the Mediterranean countries, N. Africa, and the Middle East. The subject has been extensively reviewed by Mager *et al.* (1980).

2-Thiouracil (11), a well-known antithyroid compound, has been reported to occur in various species of *Brassica* and has been cited (Forsyth, 1967) as one of the active principles in the hypothyroidism of lambs and calves which follows feeding pregnant cows and ewes on kale (*Brassica oleracea*). 2-Thiouracil is also a well authenticated example of a free pyrimidine as an effective antiviral agent; it is known to inhibit a number of plant viruses (Brockman and Anderson, 1963).

Another pharmacologically active pyrimidine, toxopyrimidine, arises as a result of the enzymic hydrolysis of thiamine (vitamin B_1). This

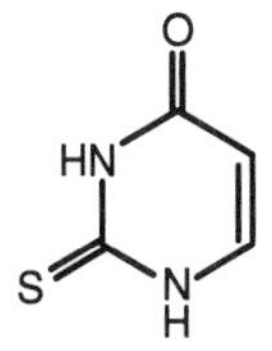

(11)

process splits the molecule between the methylene bridge and the quaternary *N*-atom, so separating and liberating the pyrimidine and thiazole moieties (Fig. 5.24). The microbial enzyme catalysing this hydrolysis is thiaminase II and the pyrimidine product (Fig. 5.24) is 2-methyl-4-amino-5-hydroxymethylpyrimidine (toxopyrimidine). Toxopyr-

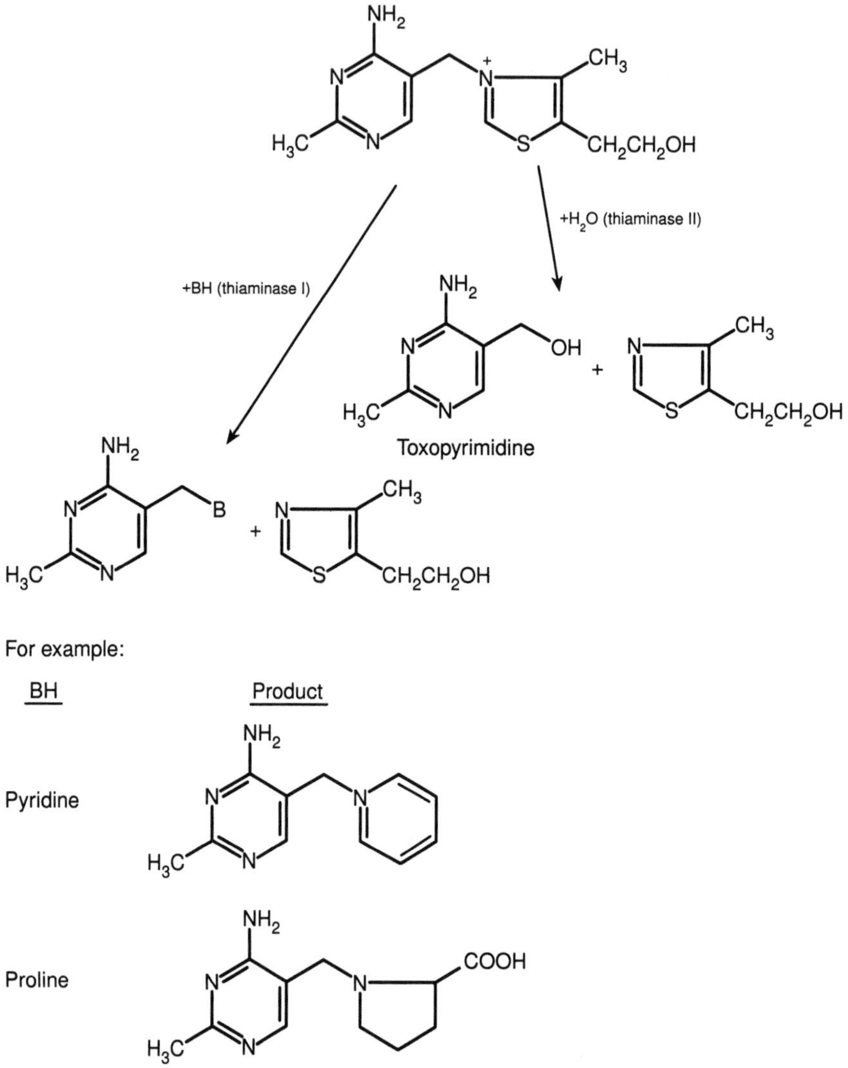

NH2 CH3 N + N H3C N S CH2CH2OH

+H2O (thiaminase II)

+BH (thiaminase I)

Toxopyrimidine

For example:

BH | Product

Pyridine

Proline

COOH

Fig. 5.24 Formation of toxopyrimidine and other pyrimidines from thiamine. Thiaminase I catalyses a base-exchange reaction involving a nucleophilic displacement on the methylene group of the pyrimidine moiety by a base BH. Thiaminase II catalyses the hydrolysis of thiamine at the bond between the methylene bridge and the quaternary nitrogen by a molecule of water (Evans *et al.*, 1982).

imidine is a structural analogue and effective antagonist of pyridoxal and in rats and mice it induces running fits which can be alleviated by administration of pyridoxal. A related enzyme, thiaminase I, which occurs in some species of fish and shellfish, and in ferns and bacteria, catalyses a fission of the thiamine molecule in a base-exchange reaction (Fig. 5.24) involving a nucleophilic displacement on the methylene group of the pyrimidine moiety by a base (BH) such as pyridine. A number of other heterocyclic bases will act as the co-substrate for this reaction and the product in each case is a substituted pyrimidine which can also be regarded as a thiamine analogue.

There are a number of pyrimidine antibiotics. Gougerotin, for example (12) shows a wide spectrum of antibacterial activity. It is an aminoacyl-tRNA analogue, which blocks protein biosynthesis by acting as an inhibitor of peptide chain elongation. Blasticidin S (13) and Amicetin A (14) also function as antibiotics by blocking protein biosynthesis but they have rather a different selectivity. Plicacetin

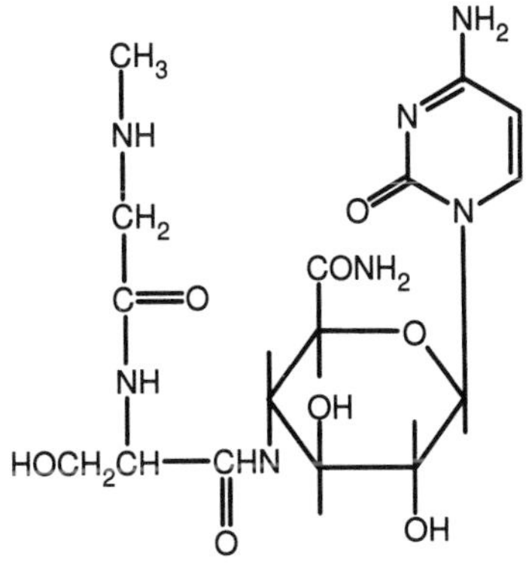

CH_3 NH CH_2 C=O NH $HOCH_2CH$—CHN O NH_2 N O N $CONH_2$ O OH OH

(12)

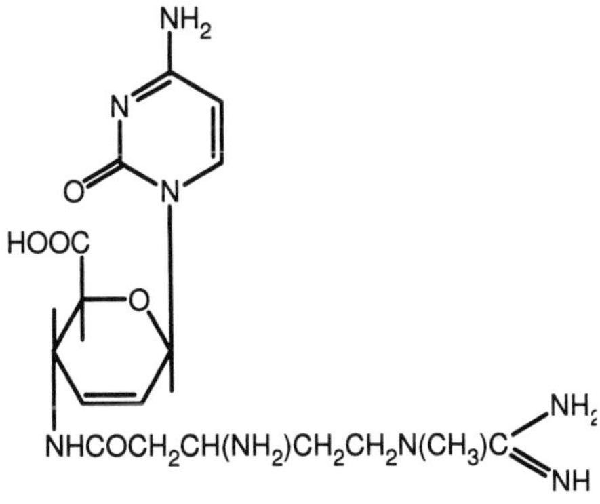

NH_2 N O N HOOC O $NHCOCH_2CH(NH_2)CH_2CH_2N(CH_3)C$ NH_2 NH

(13)

CH_3 $HOCH_2$—C—CONH— —CONH NH_2 N O N CH_3 O OH $(CH_3)_2N$ OH O CH_3 O

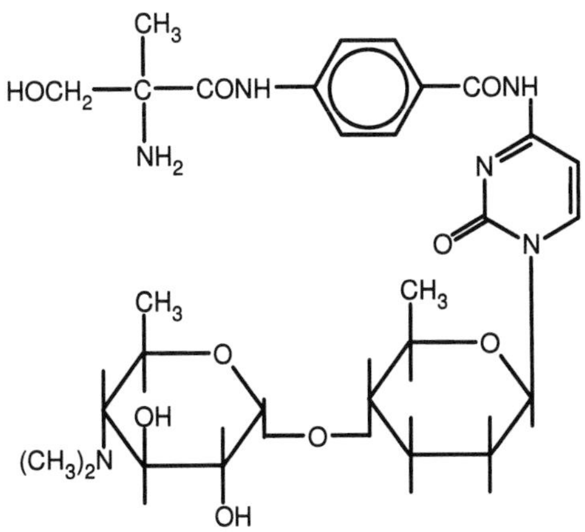

(14)

(15), also known as Amicetin B, and Bamicetin (16) which has been tentatively identified as a demethylated form of Plicacetin, have antibiotic properties, too, although these do not appear to have been studied in any detail. This group of pyrimidine antibiotics was reviewed by Fox (1966) and Korzybski *et al.* (1967).

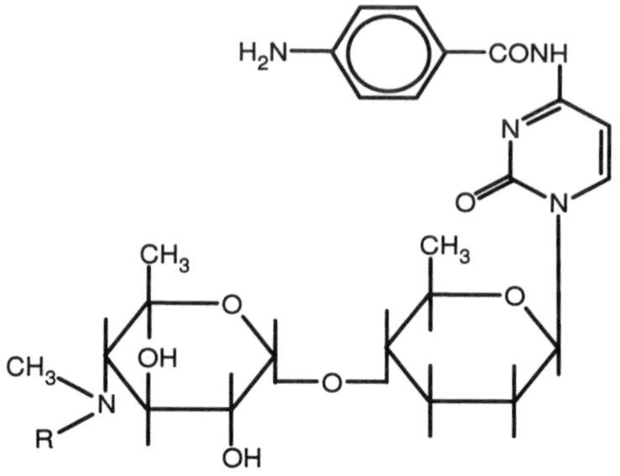

Amicetin B (Plicacetin), R = CH_3;
Bamicetin, R = H (tentative structure)

(15,16)

Another family of pyrimidine-derived antibiotics are the polyoxins, of which Polyoxin A (17) is an example. Members of this group of peptidyl-pyrimidine nucleosides are particularly effective against fungal

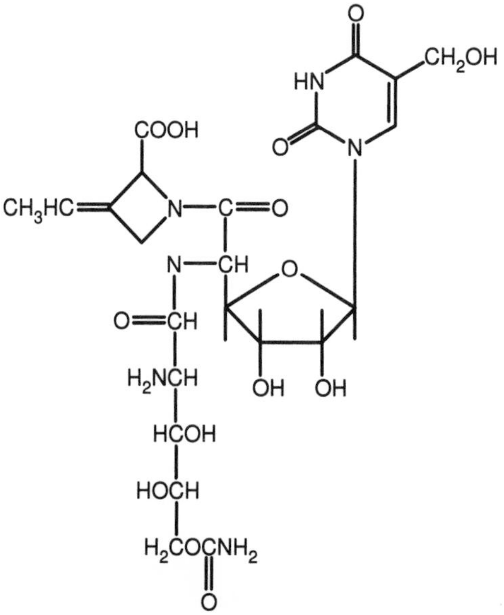

(17)

pathogens and function as structural analogues and antagonists of UDP-*N*-acetylglucosamine which is a precursor of chitin needed for cell wall synthesis.

5.7 SYNTHETIC PYRIMIDINES OF PHARMACEUTICAL INTEREST

In addition to the naturally occurring pyrimidines with biological activity, a variety of synthetic pyrimidines are of clinical importance and are used in treating a range of diseases. In cancer chemotherapy, for example, 5-fluorouracil (18), 5-fluorouridine (19) and 5-fluoro-2′-deoxyuridine (20) have been widely used as effective inhibitors of tumour nucleic acid synthesis. These compounds undergo *in vivo* conversion into the corresponding nucleotides and are particularly potent inhibitors of thymidylate synthase, the enzyme catalysing the 5-methylation of 2′-deoxyuridine 5′-monophosphate (dUMP) to 2′-deoxythymidine 5′-monophosphate (dTMP), a precursor of DNA. Unfortunately, the 5′-fluorinated uracils are not completely selective in their effects and can get incorporated into messenger RNA both in cancer and normal cells thereby interfering with physiological as well as pathological processes. This is an example of how more detailed knowledge of the active site of an enzyme, in this case thymidylate synthase, would greatly facilitate design of highly specific inhibitors for clinical application. 5-Iodouracil

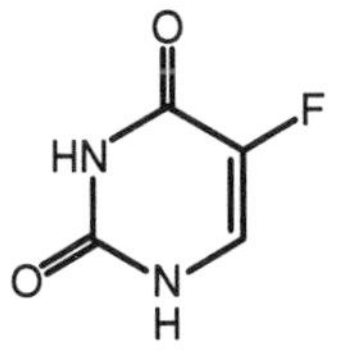

O
F
HN
O
N
H

(18)

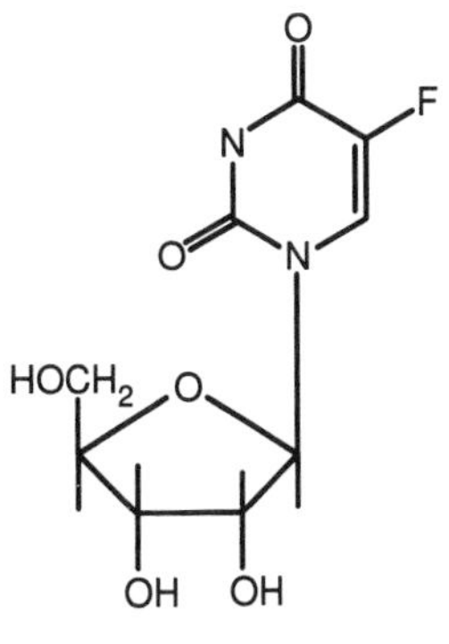

O
F
N
O
N
HOCH$_2$
O
OH OH

(19)

O
F
N
O
N
HOCH$_2$
O
H
OH H

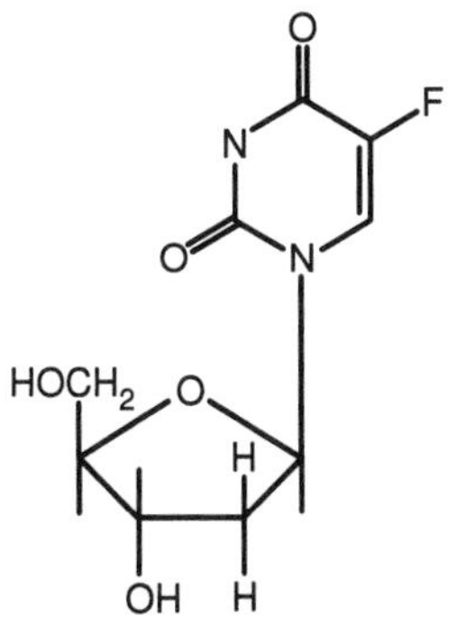

(20)

also has antitumour activity but is short-acting and less useful clinically than the fluoro-derivative. 5-Iodo 2′-deoxyuridine is, however, of substantial value as an antiviral agent in the treatment of herpes simplex infections, especially of the corneal epithelium. It is also active against vaccinia viruses. The mode of action of the iodo-derivatives is different from that of the fluoro-compounds; instead of inhibiting thymidylate synthase they act as thymidine analogues and are incorporated into DNA. 5-Bromo 2′-deoxyuridine is also an excellent thymidine analogue. It is incorporated into DNA and exhibits mutagenicity. Since the bromodeoxy UMP residues in DNA debrominate readily when DNA containing it is irradiated with UV or near-UV light, bromo-deoxy UMP also functions as a radiosensitizing agent by generating free radicals which damage the DNA structure.

6-Azauridine (21) is an effective inhibitor of OMP decarboxylase, the enzyme which catalyses the decarboxylation of orotidine 5′-monophosphate in the orotate pathway of pyrimidine biosynthesis (Fig. 5.8). In consequence, it blocks the *de novo* synthesis of pyrimidine nucleotides and thus interferes with nucleic acid formation. Not surprisingly, therefore, this uridine derivative has also found clinical application in cancer chemotherapy. Another pyrimidine nucleoside analogue that has also been used in cancer treatment is cytosine arabinoside, the arabinose analogue of cytidine; this also exhibits antiviral activity and like 5-Iodo 2′-deoxyuridine, is used in treating infections by DNA viruses such as herpes simplex. Its mode of action has been identified as inhibition of the reduction of cytidine 5′-monophosphate to the corresponding 2′-deoxy 5′-monophosphate, i.e. blocking the formation of deoxyCMP from CMP. DeoxyCMP is a precursor of DNA.

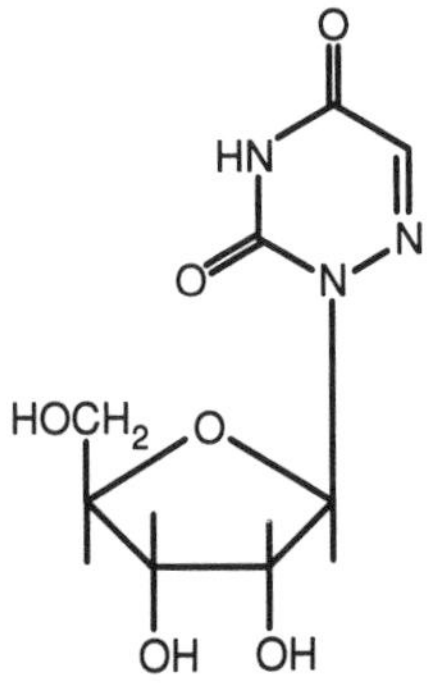

(21)

Two pyrimidine nucleoside analogues of particular topical interest in pharmacology are 3′-azido 2′-deoxythymidine (AZT; Retrovir) (22) and 2′,3′-dideoxycytidine (DDC) (23). Both of these compounds are used clinically in the management of acquired immunodeficiency syndrome

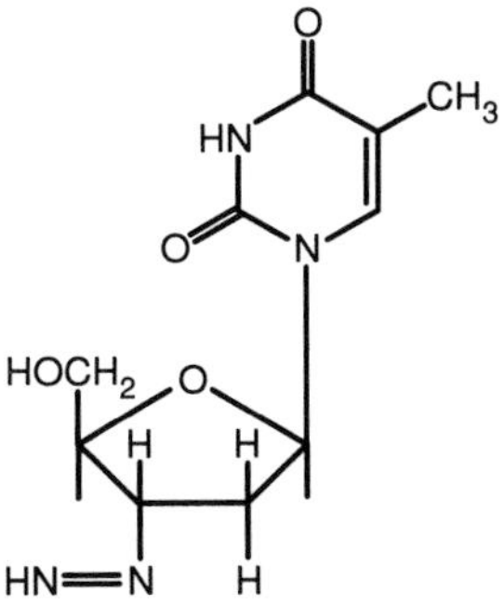

(22)

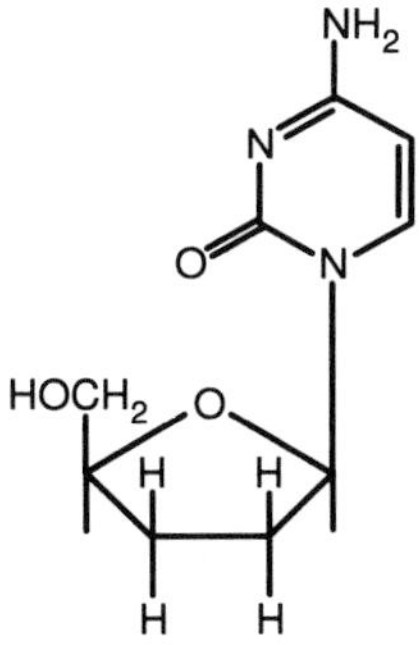

(23)

(AIDS). AZT is an analogue of 2′-deoxythymidine (24); it is phosphorylated by thymidine kinase to yield the azido-analogue of 2′-deoxyTMP (i.e. 3′-azido 2′-deoxythymidine 5′-monophosphate). After further phosphorylation to the 5′-triphosphate, this enters virus-directed DNA synthesis, catalysed by reverse transcriptase. However, as the nucleotide analogue has no 3′-hydroxyl group, it prevents extension of the growing DNA chain. Unfortunately, although AZT is highly selective for viral nucleic acid synthesis, it still shows some toxicity towards bone marrow cells. In a similar way, dideoxycytidine functions as an analogue of 2′-deoxycytidine (25), the 5′-triphosphate of which is another precursor of DNA.

Trimethoprim (26) is widely used clinically in the treatment both of

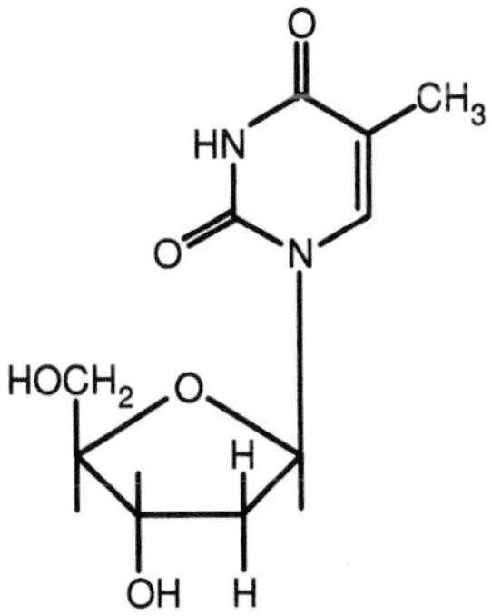

(24)

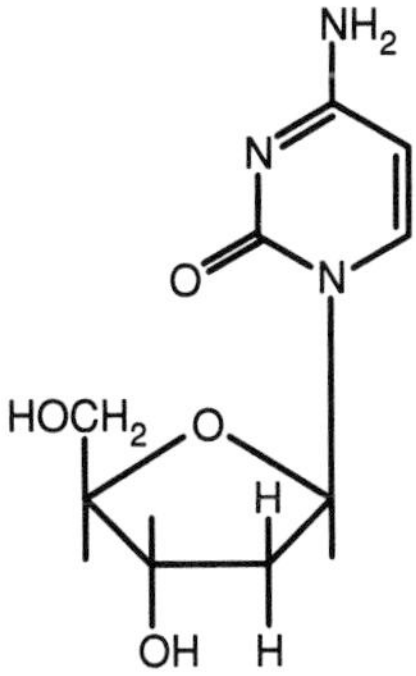

(25)

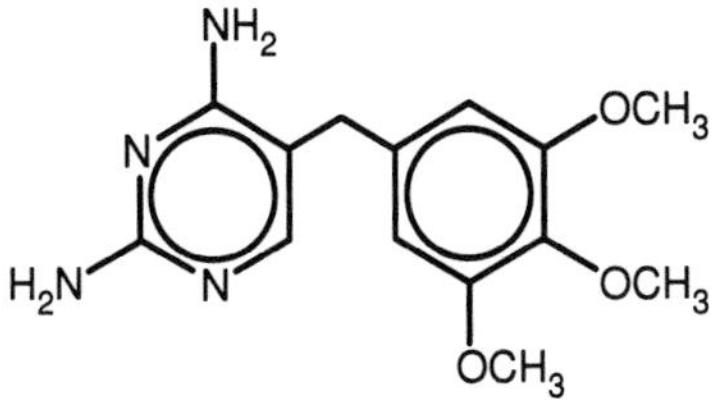

(26)

bacterial infections and of some forms of malaria. It functions as a competitive inhibitor of the enzyme dihydrofolate reductase which reduces the vitamin folic acid (27) in a two-step process, to its coenzyme form tetrahydrofolic acid. The effectiveness of Trimethoprim in chemotherapy is attributable to its selectivity. Whereas it is a strong inhibitor of the microbial enzyme, it is only a weak inhibitor of vertebrate dihydrofolate reductase.

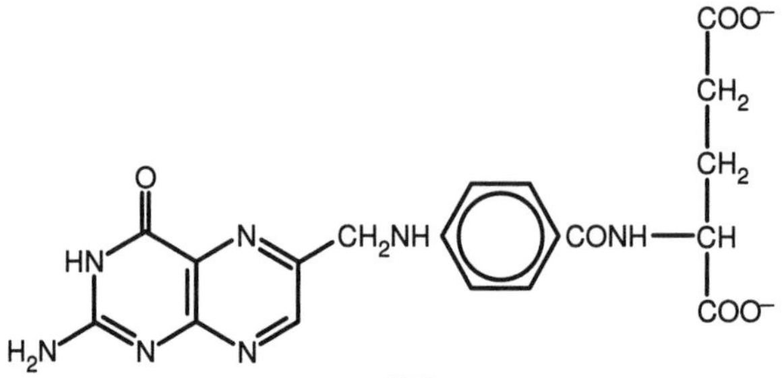

(27)

A wide range of other pyrimidines have been synthesized and tested for antimalarial activity. Of these, one of the most effective is Daraprim (2,4-diamino-5-(*p*-chlorophenyl)-6-ethylpyrimidine) (28).

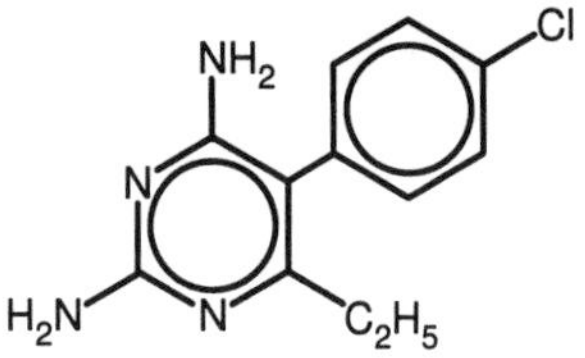

(28)

Pharmacological interest in diuretics has resulted in the synthesis of a number of pyrimidines exhibiting clinically useful activity. One of the most successful of these has been amisometradine (Rolicton) (29). They appear to exert their diuretic effects by specifically inhibiting tubular reabsorption of Na^+ and K^+ ions without either increasing the glomerula filtration rate or affecting factors regulating the acid-base balance.

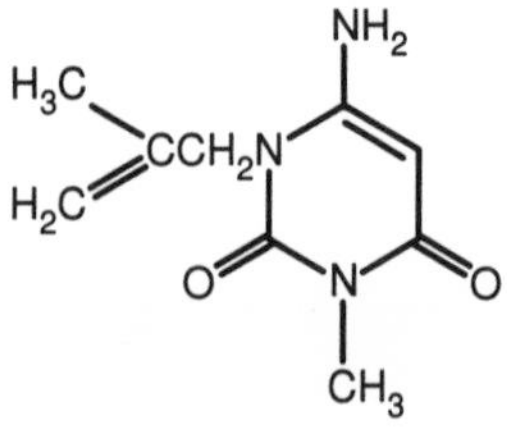

(29)

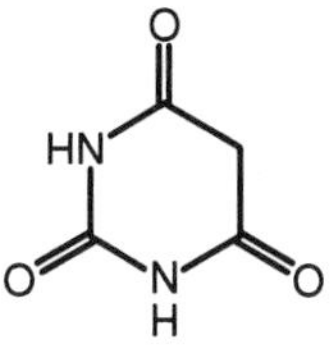

(30)

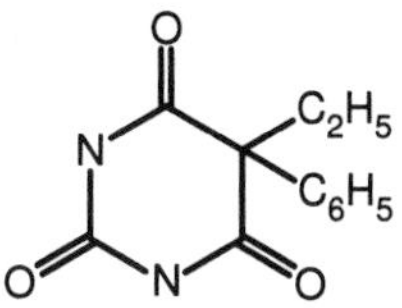

(31)

A family of pyrimidine derivatives of long standing pharmaceutical interest and clinical application are the derivatives of barbituric acid (30). Barbituric acid was discovered over 100 years ago by Bäyer and the first hypnotic barbiturate was introduced into medicine in 1903, by Fischer and von Mehring. Since then, more than 2,500 barbiturates have been synthesized but only a few of these are now in common pharmaceutical use, mainly as sedatives, hypnotics and anticonvulsants. Studies of the relationship of structure to pharmacological activity have shown that to be active, barbiturates must possess two lipophilic substituents at the 5-position and each of these must have at least 2 carbon atoms (see, e.g. phenobarbitol, 31). In general, the longer these substituents, the more pronounced are the hypnotic properties. Linked to this is the fact that hypnotic activity and its duration are related to the lipid solubility of the compound. Despite the interest in barbiturates, relatively little is known of their mode of action. They tend to be hydroxylated, as part of the detoxification process, by liver cytochromes P-450 and they act as inducers of the synthesis of these haem proteins. Some barbiturates, e.g. amytal, are known to be specific inhibitors of the respiratory electron transport chain.

Alloxan (32) is another pyrimidine derivative exhibiting interesting biological activity. Structurally, it is closely related to barbituric acid but its activity lays in a totally different direction. It specifically attacks the β-cells of the islets of Langerhans, the endocrine tissue of the pancreas responsible for the synthesis of insulin, and has consequently been used in diabetes research to induce this condition in experimental animals. Its mode of action is unknown and the relationship between the structure of alloxan and its apparent cytospecificity poses yet another intriguing but unsolved problem in heterocyclic biochemistry.

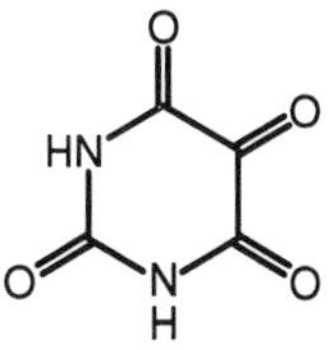

(32)

5.8 PYRIMIDINES IN HORTICULTURAL AND AGRICULTURAL USE

Several pyrimidines have useful herbicidal activity, in particular the substituted uracils Bromacil (33), Isocil (34) and Terbacil (35) have found wide application in agriculture as highly effective broad-spectrum herbicides with low mammalian toxicity. These compounds are direct inhibitors of the photosynthetic apparatus and are believed to exert this effect by blocking photosynthetic electron transport within the chloroplast. Three pyrimidine fungicides, Dimethirimol (36), Ethirimol (37) and Bupirimate (38) are in common horticultural use against powdery mildew of cucumbers and other crops. The mode of action of these compounds is believed to be as inhibitors of purine biosynthesis, or possibly as pyridoxal antagonists. Representatives of the pyrimidines are also found amongst the commonly used horticultural insecticides, e.g. Pirimicarb (39) and Diazinon (40) where it is suggested they act as acetylcholinesterase inhibitors. Castrix (41) and its 4-methylamino-analogue, are widely used rodenticides.

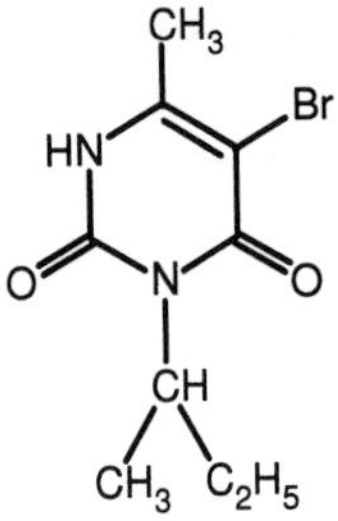

(33) 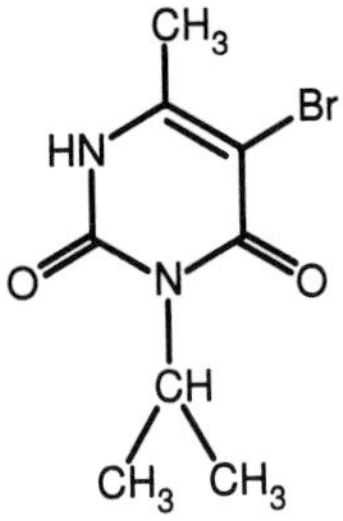 (34) 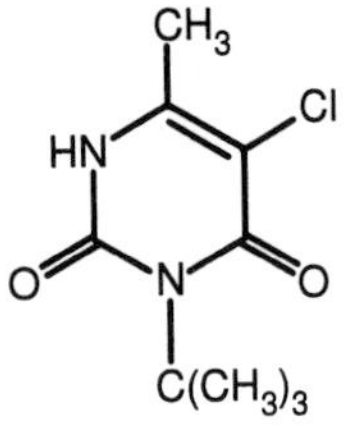 (35)

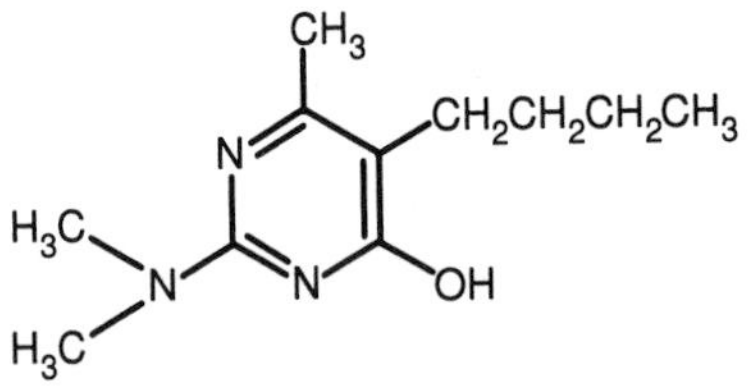

(36) 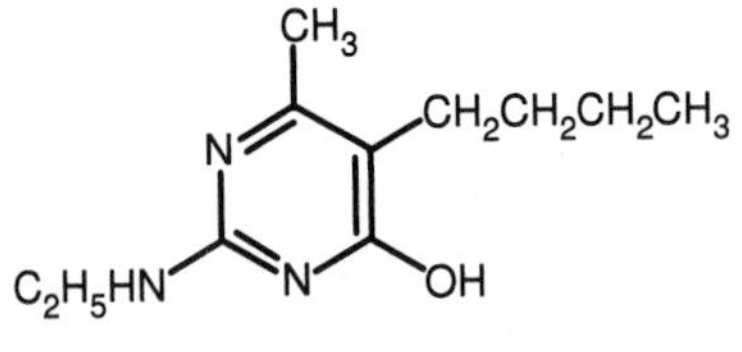 (37)

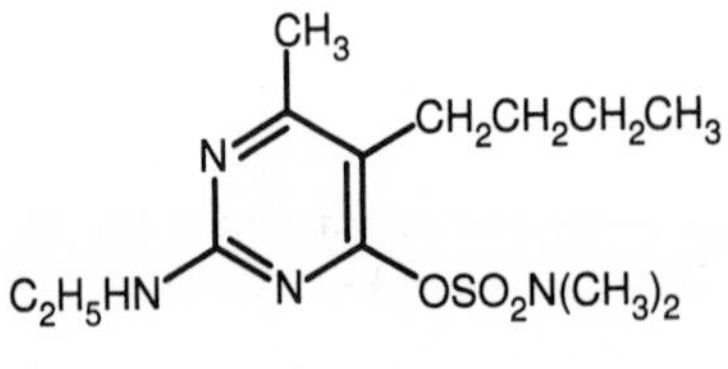

(38) 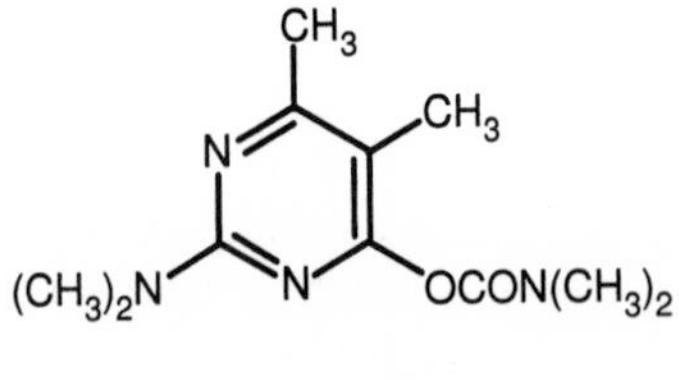 (39)

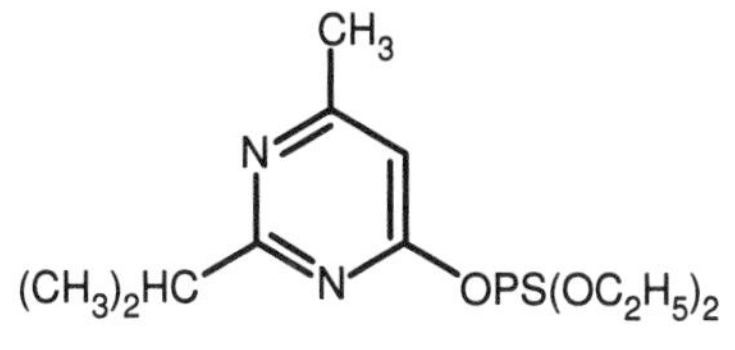

(40) 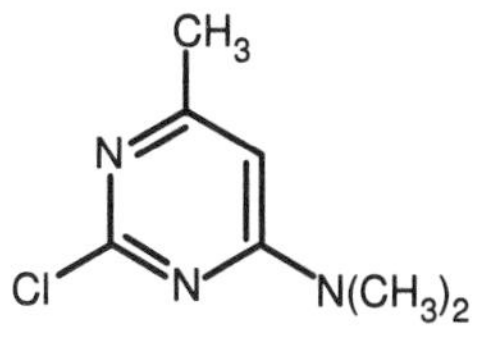 (41)

5.9 PYRIMIDINE CHEMISTRY AND BIOCHEMISTRY, CONCLUSIONS

Pyrimidines are not only components of RNA and DNA, and hence of the information storage, retrieval and transfer system of all living systems, but in their own right they are coenzymes, precursors of coenzymes, regulators and integrators of metabolism. The only other compounds involved to a comparable extent in these central areas of metabolism are the purines but they too could be considered to be pyrimidine derivatives, i.e. fused pyrimidines. What is there then about the properties of pyrimidines that equips them for these central metabolic roles? Consideration of the chemistry of the pyrimidine ring system yields surprisingly few clues. Pyrimidines are typical aromatic heterocycles with their annular 1,3-*N*-atoms reinforcing each other in causing a marked π-electron deficiency at the 2, 4 and 6 position. Due to induction, the 5-position is also probably slightly electron deficient although on the whole it behaves as a normal benzenoid position. Because of their π-electron deficiency, the 2, 4 and 6 positions are readily attacked by nucleophiles; if electrophilic substitution occurs at all, it is at the benzenoid 5-position. This overall view of the chemistry of the pyrimidine ring is however substantially modified in the naturally occurring pyrimidines by virtue of the fact that, almost without exception, they have electron releasing substituents such as hydroxyl or amino groups which counteract the π-electron deficiency. This makes the pyrimidine ring more like a typical aromatic ring, facilitating electrophilic substitution while deactivating the 2, 4 and 6 positions to nucleophilic attack.

The relationship between the chemical properties of pyrimidines and their biological functions is more easily rationalized in the case of the nucleic acids. DNA consists of a duplex of two helically interacting strands which split into individual strands during cell division. Each of these is then used as a template for the construction of the complementary strand. The accuracy of this process, essential as it is to the functioning of the biological information system, is ensured by the structure of the component pyrimidine and purine bases. This is because of specific base-pairing between the complementary strands (Fig. 5.25) which, in turn, is a function of the strong hydrogen bonding that is possible

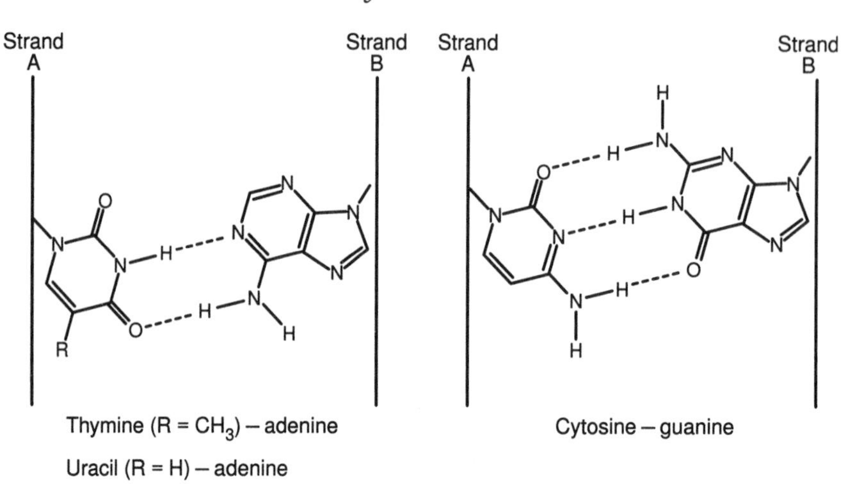

Fig. 5.25 Base-pairing between complementary strands of the DNA duplex.

between thymine (or uracil) and adenosine on the one hand, and between cytosine and guanine on the other. In practice this means that each time a base is added to the newly forming chain, it must be capable of strong hydrogen bonding with its counterpart on the template, otherwise it cannot be locked into position in the duplex. A more detailed account of these processes can be found in the books by Adams *et al.* (1992) and Blackburn and Gait (1990).

Consideration of the transport functions of UDP and CDP poses a number of interesting biochemical questions. For example, what is it about the structure and properties of UDP that especially mark it out for the role of glycosyl transporter? There are numerous examples of UDP-hexoses including UDP-glucose, UDP-galactose, and a wide range of related compounds such as the UDP-uronic acids (e.g. UDP-glucuronic acid and UDP-galacturonic acid) and UDP-aminosugars like UDP-*N*-acetylglucosamine and UDP-*N*-acetylgalactosamine. UDP-pentoses also occur but the range is much narrower; UDP-xylose and UDP-arabinose are examples. Another question that arises in this field of UDP-sugar biochemistry is how is it that the sugar residues transported in this way are, almost without exception, aldoses. As described in Section 5.6, a UDP-ketose, namely UDP-fructose, has been found in germinating peas and in tubers of Jerusalem artichoke but its biochemical function, if any, remains obscure.

Just as UDP seems to have been selected in nature for a glycosyl carrier role, so too has CDP seemingly acquired the specific biochemical function of transporting alcohol residues. Examples include CDP-ribitol, CDP-choline, CDP-ethanolamine and the CDP-diacylglycerols. Again this poses a question of fundamental importance. What is the underlying chemistry that separates these two pyrimidine nucleoside diphosphates, UDP and CDP, into their respective transporting functions? The only difference in the structure is that UDP has a hydroxyl group at position 4 which is replaced by an amino group in CDP (Table 5.2). Whereas this may seem a relatively small difference to have such a noticeable effect on function, it should be borne in mind that 4-aminopterin, a potent antifolate drug, has only a similar small difference in the structure of its pyrimidine ring (43) to that of folate itself (42) yet the effect is profound.

(42)

(43)

REFERENCES

Ahmmad, M.A.S., Maskall, C.S. and Brown, E.G. (1984) Partial purification and properties of willardiine and isowillardiine synthase activity from *Pisum sativum*. *Phytochemistry*, **23**, 265–270.

Al-Baldawi, N.F. and Brown, E.G. (1983) Accumulation of 5-ribosyluracil (pseudouridine) within the tissues of *Phaseolus vulgaris*. *Phytochemistry*, **22**, 419–421.

Al-Baldawi, N.F. and Brown, E.G. (1983) Metabolism of [6-^{14}C]orotate by shoots of *Pisum sativum*, *Phaseolus vulgaris*, and *Lathyrus tingitanus*. *Phytochemistry*, **22**, 1925–1928.

Ascoli, A. (1900) *Hoppe-Seyler's Z. Physiol. Chem.*, **31**, 161–164.

Ashworth, T.S., Brown, E.G. and Roberts, F.M. (1972) Biosynthesis of willardiine and isowillardiine in germinating pea seeds and seedlings. *Biochem. J.*, **129**, 897–905.

Bell, E.A. (1963) A new amino acid, γ-hydroxyhomoarginine in *Lathyrus*. *Nature*, **199**, 70–71.
Bell, E.A. and Przybylska, J. (1965) The origin and site of synthesis of the pyrimidine ring in the amino acid lathyrine. *Biochem. J.*, **94**, 35P.
Bendich, A. and Clements, G.C. (1953) Revision of the structural formulation of vicine and its pyrimidine aglycone divicine. *Biochim. Biophys. Acta*, **12**, 462–477.
Bieleski, R.L. (1964) The problem of halting enzyme action when extracting plant tissues. *Analyt. Biochem.*, **9**, 431.
Bien, S., Salemnik, G., Zamir, L. and Rosenblum, M. (1968) *J. Chem. Soc.* (C), **446**.
Brockman, R.W. and Anderson, E.P. (1963) in *Metabolic Inhibitors*, (eds M. Rochstel and J.H. Quastel), Academic Press, New York and London.
Brown, E.G. (1991) Purines, pyrimidines, nucleosides and nucleotides. *Methods in Plant Biochemistry*, **5**, 53–90, Academic Press, London.
Brown, E.G. (1996) 2-Amino-4-carboxypyrimidine in seeds of *Lathyrus tingitanus*. *Phytochemistry*, **42**, 61–62.
Brown, E.G. and Al-Baldawi, N.F. (1977) Biosynthesis of the pyrimidine amino acid lathyrine by *Lathyrus tingitanus*. *Biochem. J.*, **164**, 589–594.
Brown, E.G. and Mangat, B.S. (1967) UDP-fructose in germinating pea seeds. *Biochim. Biophys. Acta*, **148**, 350–355.
Brown, E.G. and Mohamad, J. (1990) Biosynthesis of lathyrine; a novel synthase activity. *Phytochemistry*, **29**, 3117–3121.
Brown, E.G. and Roberts, F.M. (1972) Formation of vicine and convicine by *Vicia faba*. *Phytochemistry*, **11**, 3203–3206.
Brown, E.G. and Turan, Y. (1996) Lathyrine biosynthesis and the origin of the precursor 2-amino-4-carboxypyrimidine in *Lathyrus tingitanus*. *Phytochemistry*, **43**, 1029–1031.
David, S., Estramareix, B. and Hirshfeld, H. (1967) Biosynthesis of thiamine. Formate the precursor of carbon-4 of the pyrimidine ring. *Biochim. Biophys. Acta*, **148**, 11–21.
Fischer, E. and Roeder, G. (1901) *Ber. Dtsch. Chem. Ges.*, **34**, 3751.
Forsyth, A.A. (1967) Kale (*Brassica oleracea*). in *British Poisonous Plants*, HMSO, London (Reference Book 161 Ministry of Agriculture, Fisheries and Food), pp. 44–55.
Hayaishi, O. and Kornberg, A. (1952) Metabolism of cytosine, thymine, uracil and barbituric acid by bacterial enzymes. *J. Biol. Chem.*, **197**, 717–732.
Hotchkiss, R.D. (1948) *J. Biol. Chem.*, **175**, 315–332.
Knowles, P.E., Marsh, D. and Rattle, H.W.E. (1976) *Magnetic Resonance of Biomolecules*, John Wiley & Sons, Chichester.
Kossel, A. and Neumann, A. (1893) Ueber das Thymin, ein Spaltungsprodukt der Nukleïnsäure. *Ber. Dtsch. Chem. Ges.*, **26**, 2753.
Kossel, A. and Neumann, A. (1894) Darstellung und Spaltungsprodukte der Nucleïnsäure (Adenylsäure). *Ber. Dtsch. Chem. Ges.*, **27**, 2215–2222.
Kossel, A. and Steudel, H. (1903) *Hoppe-Seyler's Z. Physiol. Chem.*, **38**, 80.
London, I.M. (1961) *The Harvey Lectures*, Ser. 56 (1960-1961), Academic Press, New York, pp. 151–189.
Mager, J., Chevion, M. and Glaser, G. (1980) in *Toxic Constituents of Plant Foodstuffs*, 2nd edn, (ed I.E. Liener), Academic Press, New York and London, pp. 265–294.
Makoff, A.J. and Radford, A. (1978) *Microbiol. Rev.*, **42**, 307–328.
Newell, P.C. and Tucker, R.G. (1968) Precursors of the pyrimidine moiety of thiamine. *Biochem. J.*, **106**, 271–277; Biosynthesis of the pyrimidine moiety of thiamine, 279–287.

Nishimura, S., Taya, Y., Kuchino, Y. and Ohashi, Z. (1974) Enzymatic synthesis of 3-(3-amino-3-carboxypropyl)uridine in *E. coli* phenylalanine transfer RNA: transfer of the 3-amino-3-carboxypropyl group from S-adenosyl-methionine. *Biochem. Biophys. Res. Commun.*, **57**, 702–708.

Nowacki, E. and Nowacka, D. (1963) Biosynthesis of tingitanine. A free amino acid from Tangier pea. *Bull. Acad. Pol. Sci. Ser. Sci. Biol.*, **11**, 361–363.

O'Neil, T.D. and Naylor, A.W. (1968) Purine and pyrimidine nucleotide inhibition of carbamoyl phosphate synthetase from pea seedlings. *Biochim. Biophys. Res. Commun.*, **31**, 322–327.

O'Neil, T.D. and Naylor, A.W. (1969) Partial purification and properties of carbamoyl phosphate synthetase of Alaska Pea (*Pisum sativum*). *Biochem. J.*, **113**, 271–279.

O'Neil, T.D. and Naylor, A.W. (1976) *Plant Physiol.* **57**, 23–28.

Ong, B.L. and Jackson, J.F. (1972) Pyrimidine nucleotide biosynthesis in *Phaseolus aureus*. Enzymic aspects of the control of carbamoyl phosphate synthesis and utilization. *Biochem. J.*, **129**, 583–593.

Pollock, C.J. and Cairns, A.J. (1991) Fructan metabolism in grasses and cereals. *Ann. Rev. Plant Physiol. Plant Mol. Biol.*, **42**, 77–101.

Ritthausen, H. (1881) *J. Prakt. Chem.*, **24**, 202–220.

Ritthausen, H. and Kreusler, U. (1870) *J. Prakt. Chem.*, **2**, 333.

Samuelsson, T. and Olssen, M. (1990) Transfer RNA pseudouridine synthases in *Saccharomyces cerevisiae*. *J. Biol. Chem.*, **265**, 8782–8787.

Schram, K.H. (1989) Purines and Pyrimidines, in *Mass Spectrometry*, (ed A.M. Lawson), Walter de Gruyter, New York, Chapter 10.

Steudel, H. (1900) *Hoppe-Seyler's Z. Physiol. Chem.*, **30**, 539.

Tomlinson, R.V. (1966) Acetate as a methyl group precursor in the pyrimidine moiety of thiamine. *Biochim. Biophys. Acta*, **115**, 526–529.

Uemakso, T. and Suhadolnik, R.J. (1973) Pseudouridine: biosynthesis by *Streptoverticillicum ladakanus*. *Biochim. Biophys. Acta*, **319**, 348–353.

Umemura, Y., Nakamura, M. and Funahashi, S. (1967) *Arch. Biochem. Biophys.*, **119**, 240–252.

Waller, G.R. and Dermer, O.C. (eds) (1980) *Biochemical Applications of Mass Spectrometry*, John Wiley & Sons, Chichester, first supplementary volume.

Wasternack, C. (1982) *Encyclopaedia of Plant Physiology*, Vol. 14B, *Nucleic Acids and Proteins in Plants II*, Springer, Berlin, pp. 263–301.

Wyatt, G.R. (1950) *Nature*, **166**, 237.

Wyatt, G.R. (1951) Recognition and estimation of 5-methylcytosine in nucleic acids. *Biochem. J.*, **48**, 581.

Wyatt, G.R. and Cohen, S.S. (1952) *Nature*, **170**, 1072.

WIDER READING (BOOKS AND REVIEWS)

Adams, R.L.P., Knowler, J.T. and Leader, D.P. (1992) *The Biochemistry of the Nucleic Acids*, 11th edn, Chapman & Hall, London.

Blackburn, G.M. and Gait, M.J. (1990) *Nucleic Acids in Chemistry and Biology*, Oxford University Press, Oxford.

Brown, D.J. (1962) *The Pyrimidines*, Wiley-Interscience, New York.

Brown, D.J. (1970) *The Pyrimidines*, Supplement **1**, Wiley-Interscience, New York.

Hurst, D.T. (1980) *An Introduction to the Chemistry and Biochemistry of Pyrimidines, Purines and Pteridines*, J. Wiley, Chichester.

Jones, M.E. (1980) Pyrimidine nucleotide biosynthesis in animals: Genes, enzymes, and regulation of UMP biosynthesis, *Annual Review of Biochemistry*, **49**, 253–279.

CHAPTER 6

Purines

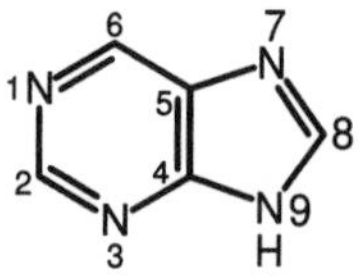

6.1 DISCOVERY AND NATURAL OCCURRENCE

The dominance of nucleic acids and molecular biology in contemporary biochemistry tends to obscure the fact that biological and chemical knowledge of the purines substantially predated their discovery as nucleic acid constituents and can be traced back over 200 years. The first of these compounds to be isolated was uric acid (1) obtained by Scheele and Bergman in 1776 from bird excreta, human urine and urinary calculi. Undoubtedly this early discovery was greatly facilitated by the relatively low solubility of uric acid and hence its tendency to crystallize easily from biological fluids and extracts. Interest in the chemistry of urinary calculi also led Marcet, some forty years later, to the discovery of xanthine (2). Guanine (3) was isolated by Magnus in 1844 from guano, hence the name, and this discovery was followed in 1850 by Scherer's isolation of hypoxanthine (4) from beef spleen. The last of the commonly occurring purines to be discovered was adenine (5) obtained by Kossel in 1885–6 from beef pancreas.

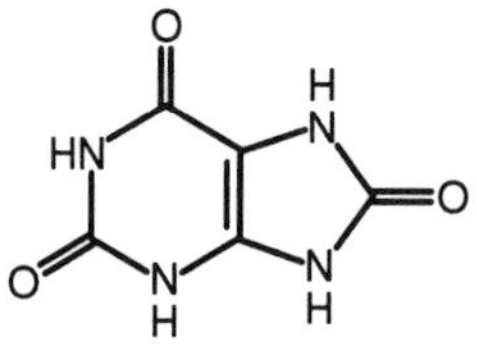

(1)

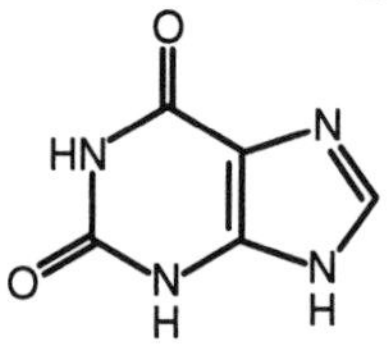

(2)

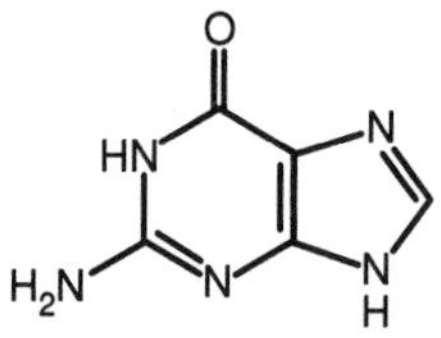

(3)

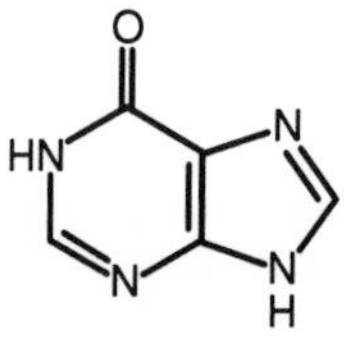

(4)

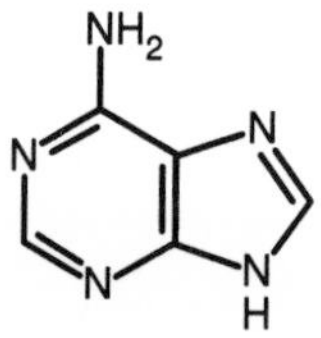

(5)

The fundamental chemistry, including structure and properties, was mainly elucidated by the classical studies of Emil Fischer during the period 1882–1906. It was Fischer who named the compounds 'purines' (*L. purum* and *uricum*) and he published prolifically on the topic, describing some 150 novel purines in the process.

Largely due to the researches of Miescher and of Kossel, and their respective collaborators, the nucleic acid story began to unfold. The early studies centred on the structure of nucleic acids as revealed by hydrolytic degradation, and led eventually to an understanding of nucleoside and nucleotide structure. One of the last major structural problems, at this level, was to locate the glycosidic bond. As nucleotides are non-reducing it was deduced that C-1 of the sugar is implicated but the major question concerned which of the purine ring nitrogens is involved. *N*-1 and *N*-3 were relatively quickly eliminated but deciding between the two remaining nitrogens, *N*-7 and *N*-9, was more difficult. In 1936, Gulland and Holiday solved the problem spectrophotometrically using *N*-7-methyladenine and *N*-9-methyladenine as model compounds. The UV-spectral properties of the nucleoside adenosine were shown to be consistent with *N*-9 substitution but not *N*-7 substitution. Todd and his colleagues confirmed this in 1949 when they accomplished the first unambiguous synthesis of adenosine (Kenner, Taylor & Todd, 1949). Nucleotides were shown to be phosphate esters of the corresponding nucleosides, the phosphoryl residue being attached in each case to one of the sugar hydroxyl groups, *i.e.* at the 2′-, 3′-, or 5′-positions. Later it became apparent that most of the nucleotides found in a free state in Nature are 5′-phosphates whereas those derived from the hydrolytic degradation of nucleic acids are the corresponding 3′-phosphates. The structures of the commonly occurring purine ribonucleosides and nucleotides are shown in Table 6.1.

In 1956, Sutherland and his co-workers isolated a novel free nucleotide which they showed to mediate the effect of adrenalin on the activation of liver phosphorylase; the compound was subsequently identified as adenosine 3′,5′-cyclic monophosphate (6), now commonly known as

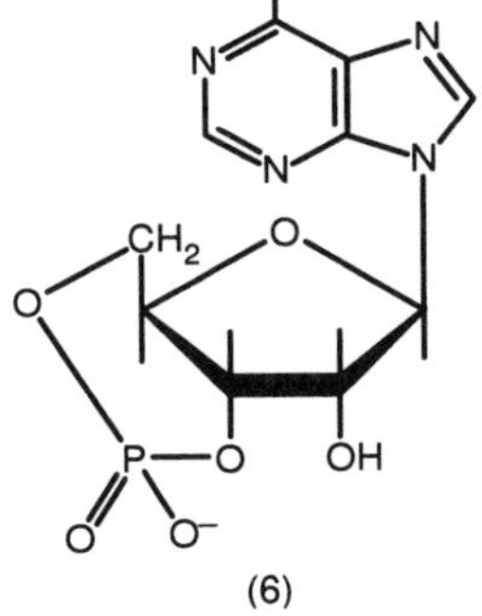

(6)

Table 6.1 The major purine ribonucleosides and ribonucleotides

Base	Ribonucleoside	Ribonucleotide
Adenine	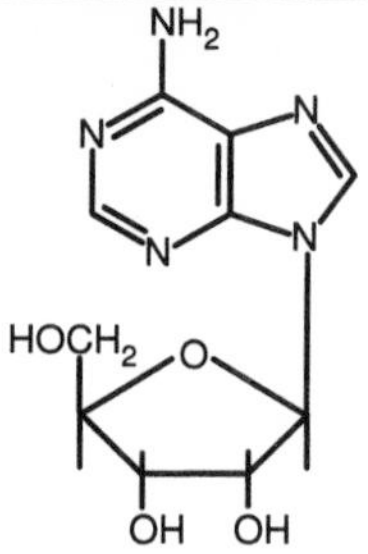 Adenosine	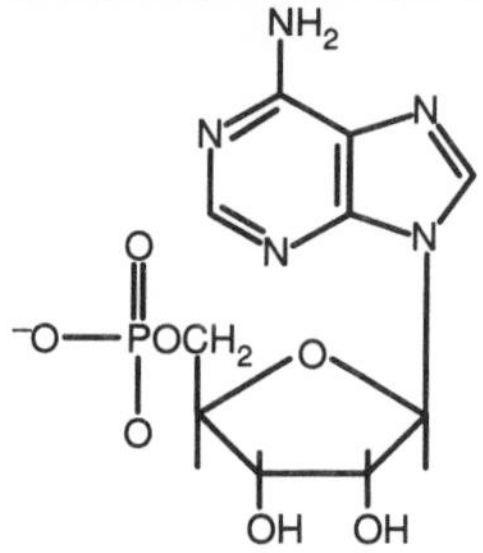 Adenosine 5'–monophosphate (AMP)
Guanine	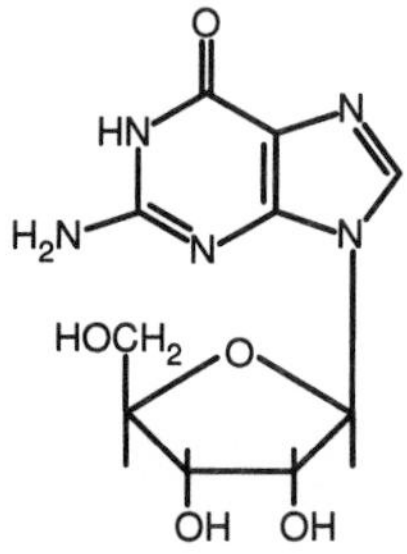 Guanosine	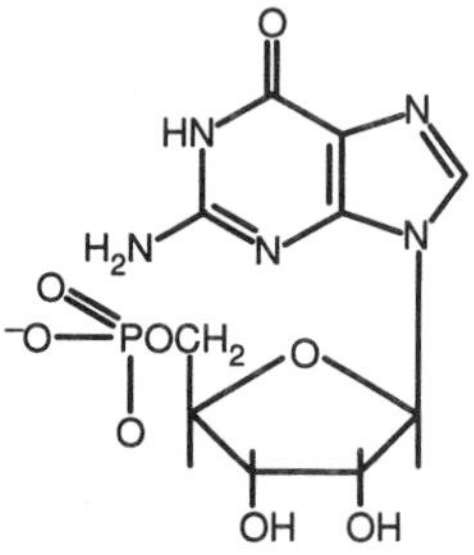 Guanosine 5'–monophosphate (GMP)
Hypoxanthine	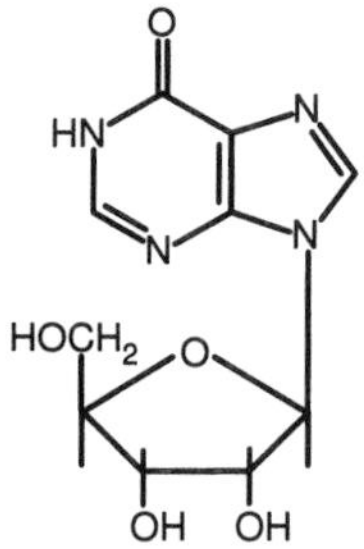 Inosine	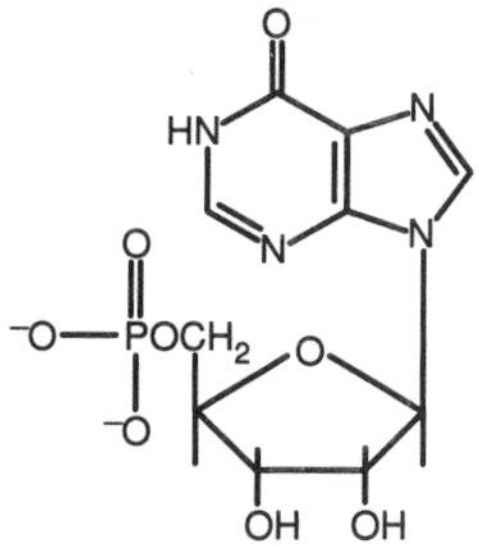 Inosine 5'–monophosphate (IMP)
Xanthine	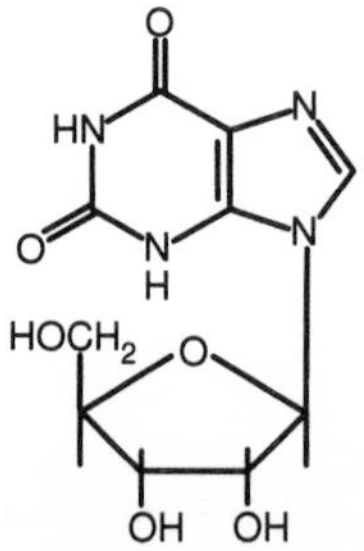 Xanthosine	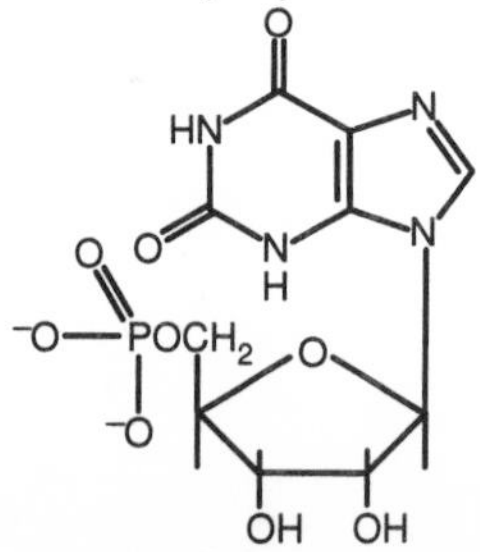 Xanthosine 5'–monophosphate (XMP)

'cyclic AMP'. A similar structure, adenosine 2′,3′-cyclic monophosphate (7), had in fact been described much earlier but is devoid of biological activity. This compound is essentially a degradation product of RNA, arising during alkali- or ribonuclease-catalysed hydrolysis.

(7)

The purine alkaloids (Fig. 6.1) are another well known group of naturally occurring purines, of interest especially in pharmacology and food technology. They are essentially plant products and mainly methylated xanthines although some methylated uric acids are also known. Caffeine (1,3,7-trimethylxanthine) is probably the best known example and it is

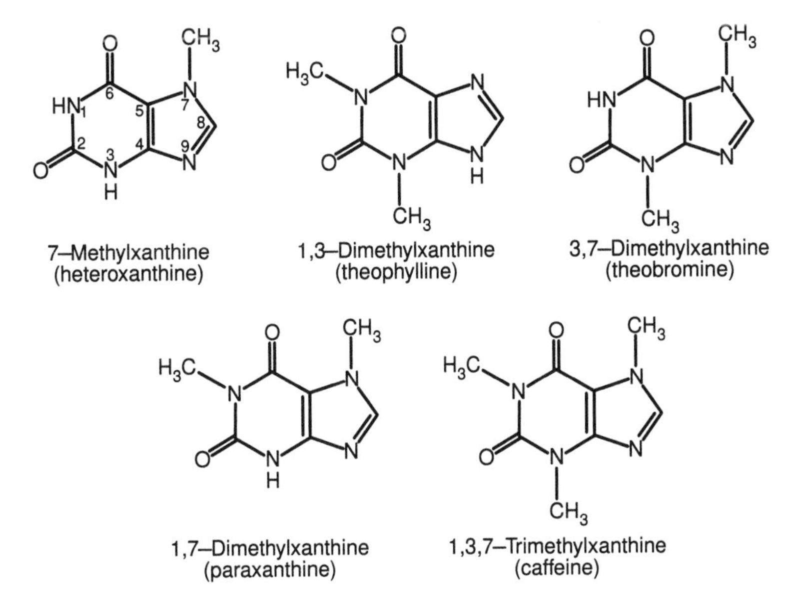

Fig. 6.1 The methylated xanthines.

said to occur in more than 60 plant species throughout the world. The most common caffeine-producing species are from the genera *Coffea, Camellia, Cola, Paullinia, Ilex,* and *Theobroma*. Without doubt, it is presence of caffeine and the related methylxanthines to which various beverages and foodstuffs, such as tea, coffee, cocoa, and various soft drinks, and chocolate, owe their long-standing international popularity. In leaves of *Camellia sinensis* (tea), more than 2% of the dry weight consists of caffeine; prepared, cured leaves may contain up to 5%. Prepared coffee beans contain about 1–2% of caffeine together with small amounts of theophylline. Prepared leaves of *Ilex paraguensis*, from which maté is made, contain 0.2–2% caffeine. Cocoa beans have a lower caffeine content (0.05–0.36%) but contain 1–3% of theobromine. Small amounts of another methylated xanthine, paraxanthine (1,7-dimethylxanthine), are also found, together with caffeine, in *Coffea arabica*. Perhaps more unexpectedly, beet juice (*Beta vulgaris*) has been shown to contain heteroxanthine (7-methylxanthine). Three methylated uric acids, namely 1,3,7,9-tetramethyluric acid, O^2,1,9-trimethyluric acid, and O^2,1,7,9-tetramethyluric acid are present in both tea and coffee. A more detailed account of the methylated purines and their metabolism, is given in a review by Suzuki, Ashihara and Waller (1992).

Plants contain another group of interesting, bioactive, purine derivatives, the cytokinins. The discovery of these compounds followed from observations made by Skoog in the late 1940s during his investigation of factors required for the growth of plant cells in tissue culture. He found that preparations of salmon sperm DNA that had been sterilized by autoclaving contained a factor that markedly stimulated plant cell division, but that this factor was absent from freshly prepared DNA samples. The active compound, given the name *kinetin*, was isolated and shown to be N^6-furfurylaminopurine (Fig. 6.2). In fact, this is not a naturally occurring compound but a rearrangement product of adenosine, formed during autoclaving. Elucidation of its structure as that of a N^6-substituted aminopurine led to the synthesis and testing of a variety of active analogues, and subsequently to the discovery of a number of naturally occurring compounds also possessing activity in stimulating plant cell division. The first of these naturally occurring compounds to be discovered, by Letham in 1963, was isolated from immature maize (*Zea mais*) kernels and given the name *zeatin*. From the 60 kg of maize extracted, 0.7 mg of zeatin was obtained. Its structure was shown to be 6-(4-hydroxy-3-methyl-*trans*-2-butenylamino)purine (Fig. 6.2). The *cis*-isomer was also found to occur naturally but is 50 times less active in the standard (tobacco callus) bioassay for cytokinin activity. Most natural cytokins have an N^6-substituent related structurally to that of zeatin (Fig. 6.2). Some of these bioactive purines occur as free ribosides and ribotides (Fig. 6.2) and in addition are components of tRNA species from a variety of biological systems including yeast, liver, spinach and

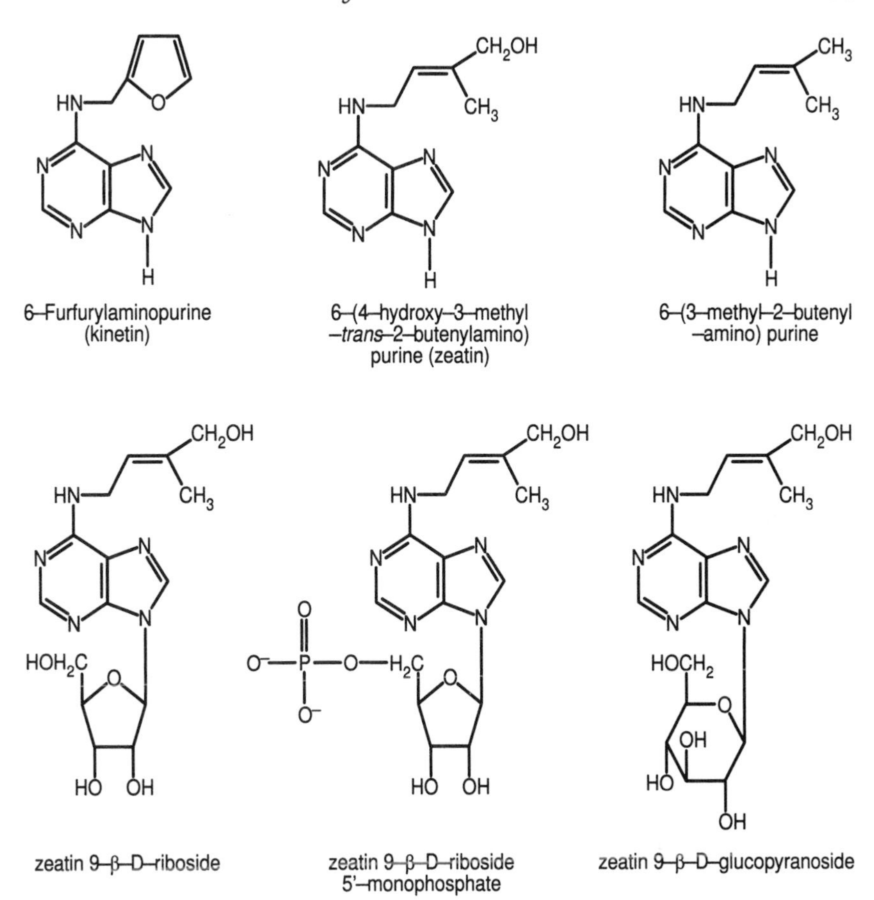

Fig. 6.2 Kinetin (6-furfurylaminopurine) and some related naturally occurring cytokinins.

peas. These tRNA components are formed by post-transcriptional modification of adenosine residues but the biological significance of this in relation to cytokinin activity is obscure.

Generally, the common purine bases, adenine and guanine, do not accumulate to any significant extent in biological tissues. Earlier reports of the widespread distribution of these compounds were, more often than not, attributable to the vigorous extraction procedures used, e.g. refluxing tissues for 2 hrs in dilute HCl or H_2SO_4, and which hydrolysed nucleic acids and nucleotides. Similarly, hypoxanthine was often found as a result of the artefactual, hydrolytic deamination of adenine. In a later period, aqueous extracts were used but these suffered from the problem of failure to inactivate hydrolytic enzymes. It is of interest that non-specific phosphatases and ribonucleases are widespread in higher plants and boiling does not inactivate them, so permitting hydro-

Table 6.2 Some of the 'minor' purine bases found in RNA

1-methyladenine	2-dimethylamino-6-hydroxypurine
2-methyladenine	7-methylguanine
6-methylaminopurine	2,2,7-trimethylguanine
6-dimethylaminopurine	hypoxanthine
6-isopentenylaminopurine	1-methylhypoxanthine
1-methylguanine	xanthine

lysis of nucleic acids and nucleotides to release nucleosides and free purine bases (Brown, 1991).

It is significant that these early reports of the free occurrence of nucleic acid bases much less frequently referred to pyrimidines. The *N*-glycosidic bond of pyrimidine nucleosides is much more resistant to hydrolysis than its purine counterpart. The problem of artefacts arising from nucleic acids and nucleotides is discussed further in Chapter 5.

In addition to the common purines adenine and guanine, nucleic acids frequently contain small amounts of unusual or 'minor' purine bases, e.g. DNA contains N^6-methyladenine, and a wide range of methylated and other modified purine bases occur in RNA (Table 6.2).

6.2 CHEMICAL PROPERTIES OF BIOLOGICAL SIGNIFICANCE

The purine nucleus consists of a fused pyrimidine and imidazole ring and like pyrimidine itself, purine is a π-deficient system. There is however no counterpart on the purine molecule to *C*-5 of the pyrimidine ring and all the carbon sites are electron-deficient. This means that, with one exception, all the positions on the purine ring system are activated towards nucleophilic attack. The exception is the 8-position which, if electron-releasing substituents are present on the pyrimidine ring, is susceptible because of its location between the electron-attracting =N- and the electron-releasing -NH- of the imidazole ring. The electron-attracting character of the ring *N*-atoms of the purine ring system (=N-) activates the 2, 6, and 8 positions and any substituents at these positions. Presence of the imidazole NH group causes purine to exhibit both acidic and basic properties (pK_a 8.9, 2.4).

All the naturally occurring purines are crystalline solids. Most are amino and/or hydroxy compounds exhibiting aromaticity and tautomerism. Like the hydroxypyrimidines, at a physiological pH the hydroxypurines exist primarily in the oxo (lactam) form whereas the amino derivatives remain in the amino rather than the imino form (Fig. 6.3). Despite the physiological prevalence of the oxo-form over the hydroxy-form, the relative simplicity and unambiguity of the hydroxy designations means that these are widely used in purine nomenclature, e.g. hypoxanthine is 6-hydroxypurine.

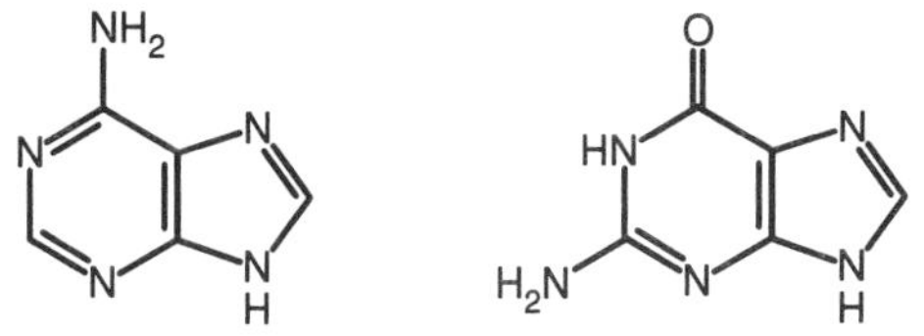

(a) (b)

Fig. 6.3 Tautomeric forms of (a) adenine, and (b) guanine, predominating under physiological conditions of pH (pH 6–7).

As the naturally occurring purines and their ribosides and ribotides all exhibit selective absorption in the ultraviolet region of the spectrum between 230 nm and 300 nm, ultraviolet absorption spectrophotometry is a particularly useful analytical tool for identifying and determining the concentrations of these compounds in biological extracts and enzymic reaction mixtures. As with the pyrimidines, however, purines exist in aqueous solution in a number of different ionic forms having individual spectral characteristics. The predominating ionic structure, and hence the spectrum obtained, is entirely pH-dependent. Consequently, in analysis care has to be taken to choose a pH at least 2 units above or below a pK_a, otherwise the spectral data obtained may represent the summation of that of several different ionic species. Use can be made of this relationship, between pH and the ionic form predominating, in order to determine pK_a values. A simple plot of absorbance at an arbitrary but fixed wavelength *versus* pH is, in effect, a dissociation curve, from which the pK_a values can be obtained. In the light of these considerations, it is surprising how many authors still quote spectral properties for aqueous solutions of purines and pyrimidines without specifying the pH of the solution. Another practical point to be remembered is that under similar conditions of pH and concentration, the spectrum of a purine riboside is virtually identical to that of the corresponding ribotide; the phosphate group is optically transparent.

Purines react readily with free radicals such as the hydroxy radicals formed by irradiation of aqueous solutions. The attack is preferentially at the 6-position if this is free, and if not, it occurs at the 8-position. Under these conditions, adenine (6-aminopurine) yields 8-hydroxyadenine (8). This type of reaction has biological significance and it is

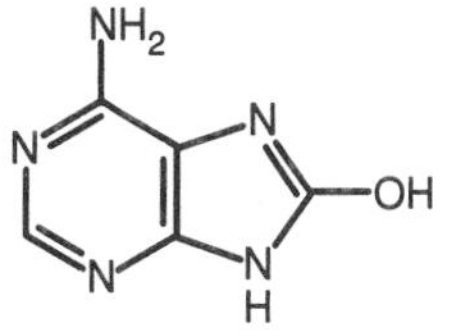

(8)

known, for example, that the intracellular generation of reactive oxygen species such as superoxide (O_2^-) induces formation of 8-hydroxyguanine (9) from guanine residues in DNA, causing mutagenesis. Superoxide can arise during oxidative metabolism by a one-electron reduction of O_2.

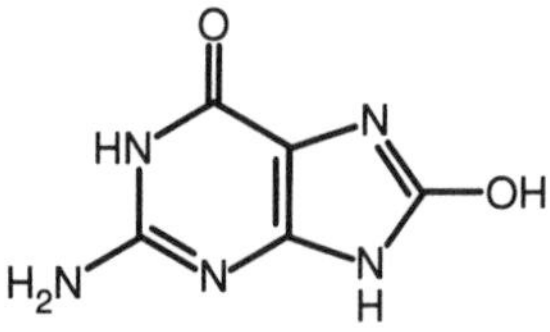

(9)

Under laboratory conditions, *N*-alkylation of purines occurs readily with a variety of alkylating agents, such as dimethyl sulphate or methyl iodide. With hydroxypurines, dialkyl rather than monoalkyl substitution takes place, e.g. xanthine treated with dimethyl sulphate in aq. KOH at 60°C yields 3,7-dimethylxanthine (theobromine). Alkylation of purine residues in DNA can occur following exposure to alkylating agents such as methyl-di-(2-chloroethyl)amine (nitrogen mustard) and bis-(2-chloroethyl)sulphide (sulphur mustard). Such compounds have been used experimentally as mutagens, carcinogens and carcinostats. Their action on DNA is complex but the *N*-7 position of guanine is preferentially attacked, followed by *N*-1 or either adenine or guanine, and finally the *O*-6 position of guanine. Alkylation of purines at *N*-7 gives rise to unstable quaternary nitrogens so that the alkylated purine may be liberated, causing a gap in the base sequence of the affected DNA molecule. It is, then, somewhat surprising to find that the formation of 7-methylguanine, the major product of DNA alkylation, does not lead to mutagenesis (Lijinsky, 1976; Lindahl, 1981; and Eadie *et al.*, 1984) although some other alkylations do (Zarbl *et al.*, 1985; Roberts, 1975). That the relationship between DNA, alkylation, and carcinogenesis is by no means a direct one is further demonstrated by the observation that good alkylating agents are often poor carcinogens. This topic is discussed further by Adams *et al.* (1992).

6.3 BIOSYNTHESIS AND INTERCONVERSION

The pathway of purine biosynthesis was largely elucidated during the decade following the end of the Second World War when isotopes first became available commercially. Studies with ^{14}C- and ^{15}N-labelled precursors, principally from the laboratory of J.M. Buchanan (Buchanan *et al.*, 1948), traced the origin of each of the constituent carbon and nitrogen atoms (Fig. 6.4) and showed that glycine was incorporated as an

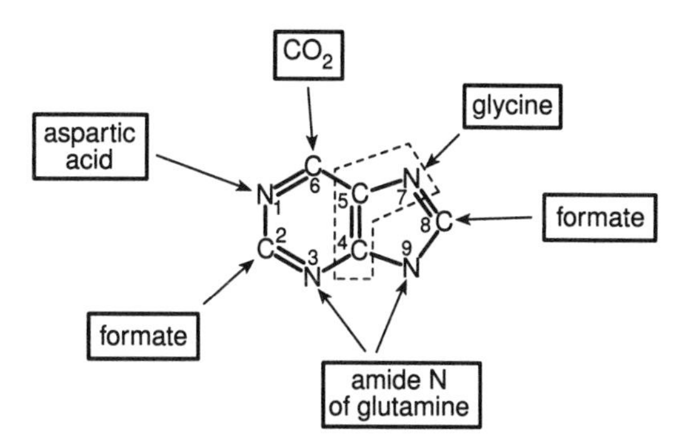

Fig. 6.4 Biosynthetic origins of the atoms comprising the purine ring system, elucidated by the use of isotopic tracers.

intact unit into the purine ring system. A major contribution to understanding of the biosynthetic process was the discovery by G.R. Greenberg (1951) that inosine monophosphate (10) is the first purine product. Having identified the simple precursors and the end-product, the search began for possible intermediates. For a time, this was hindered by the assumption that the pyrimidine ring is constructed first and it is upon it that the imidazole ring is assembled. The discovery of 5-aminoimidazole-4-carboxamide (11) in sulphonamide-inhibited bacterial cultures in which purine biosynthesis had been suppressed, led to the realization that the process, to the contrary, involved construction of a pyrimidine ring on a preformed imidazole. Subsequently, the biosynthetic intermediates were identified one after another until the complete pathway (Fig. 6.5) had been elucidated. It is of interest that the biosynthetic route starts with ribose 5-phosphate and that the whole process occurs at the ribotide level. The first-formed purine ring is that of the final product, inosine 5′-monophosphate (IMP). This contrasts markedly with pyrimidine biosynthesis during which the main part of the process proceeds at the base-level and the first-formed pyrimidine ring, that of orotic acid, then has to be raised to the nucleotide level by ribotylation to orotidine 5′-monophosphate.

IMP, the end-product of purine biosynthesis, is converted into the other biologically important purine ribonucleotides AMP and GMP by

Ribose–5–phosphate

(10)

(11)

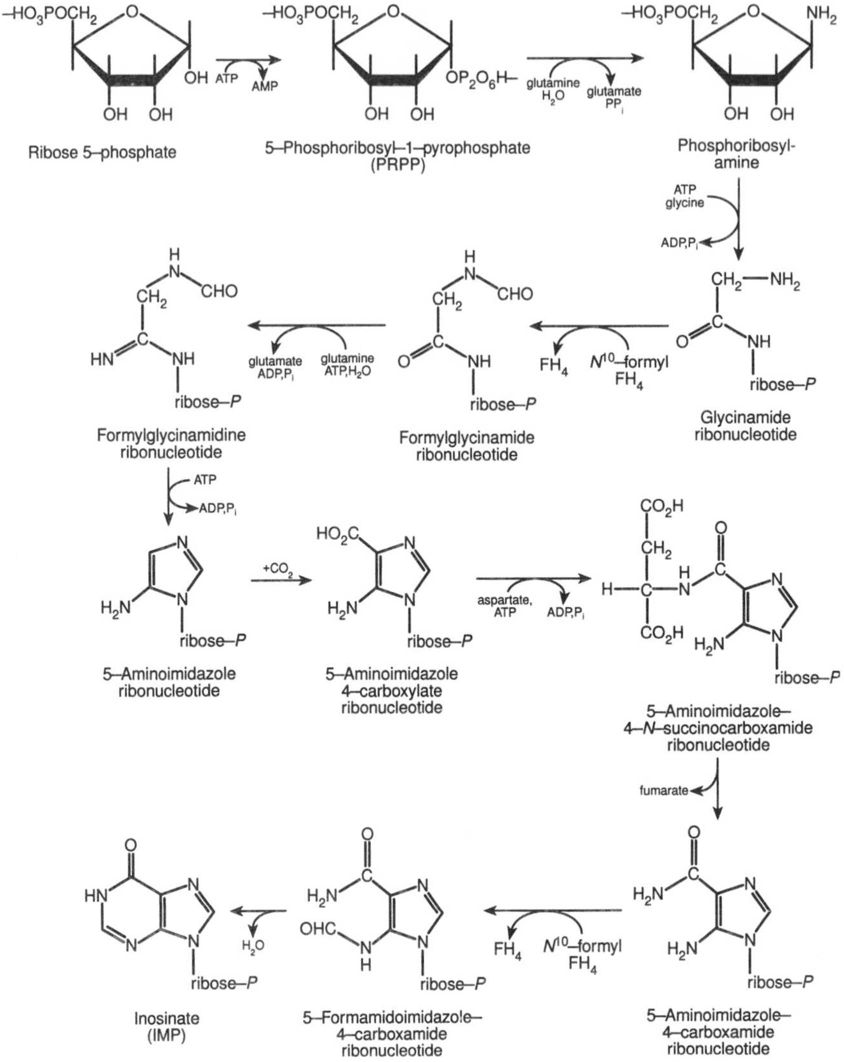

Fig. 6.5 Biosynthetic pathway for purines. Inosine 5′-monophosphate (IMP) is the end-product from which all the other purines are made in biological systems.

the series of enzymic reactions outlined in Fig. 6.6. Production of AMP from IMP involves enzymic amination with aspartate as the amino-donor and intermediate formation of adenylosuccinate. GTP is used to drive this endergonic process. For the formation of GMP, IMP must first be enzymically oxidized to xanthine 5′-monophosphate (XMP) with NAD^+ as the electron-acceptor, and then aminated to GMP. The amino donor in the second reaction is glutamine, and ATP is the energy source.

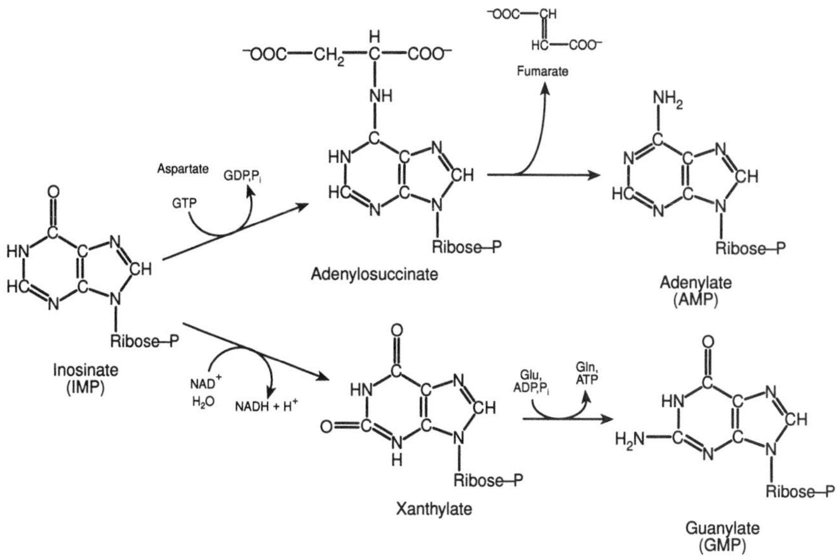

Fig. 6.6 Enzymic conversion of inosine 5′-monophosphate (IMP) into adenosine 5′-monophosphate (AMP) and guanosine 5′-monophosphate (GMP).

As with the pyrimidine ribonucleotides, reduction to 2′-deoxyribonucleotides is effected in most higher organisms and bacteria at the nucleoside 5′-diphosphate level. The enzyme catalysing the process is ribonucleoside diphosphate reductase and the reaction proceeds with retention of configuration at C-2′, implying that the mechanism does not involve replacement of the hydroxyl by a hydride ion in a S_n2 reaction. A probable mechanism is that the enzyme removes a hydrogen atom from C-3′ and that this is followed by loss of a hydroxide ion from C-2′, resulting in a resonance-stabilized radical carbonium ion. An electron pair would then be transferred from a redox thiol, accompanied in a second step by a hydrogen atom from an adjacent tyrosine residue in the enzyme-protein, to yield the deoxyribonucleotide. The ultimate source of the electrons is NADPH and the intermediate redox thiols include the tripeptide glutathione, and the proteins glutaredoxin and thioredoxin.

In addition to *de novo* purine biosynthesis, most organisms can produce nucleotides by recycling purine bases liberated during the catabolism of redundant nucleic acids or from dietary sources. A similar mechanism operates with pyrimidines. The base is directly ribotylated in an enzymic reaction with 5-phosphoribosyl-1-pyrophosphate (PRPP) as the ribotyl donor, e.g.

$$\text{adenine} + \text{PRPP} \rightarrow \text{adenosine 5 -monophosphate} + PP_i$$

Enzymes catalysing this type of reaction are known as phosphoribosyl transferases. The two concerned with purine salvage are adenine phosphoribosyl transferase and hypoxanthine-guanine phosphoribosyltransferase (HGPRT). The former is specific for adenine and the latter will only use hypoxanthine or guanine as substrate. A peculiar but rare clinical condition known to be associated with the total absence of HGPRT is the Lesch–Nyhan syndrome. First described in 1964, this genetic defect manifests itself in abnormal behaviour characterized by an unrestrainable urge to self-mutilate. Nyhan described it as like 'nail-biting with the volume turned up'. Patients are capable of gnawing off the tops of their fingers, down to the first joint, or to chew off their lower lip. A case was described of a sufferer deliberately thrusting his fingers into the rotating spokes of a hospital wheelchair in which he was being transported. How this bizarre behaviour is related to the biochemical lesion, i.e. absence of guanine-hypoxanthine phosphoribosyltransferase, remains obscure. Curiously, attempts to find an experimental animal model with which to study the Lesch–Nyhan syndrome resulted in the discovery, equally inexplicable biochemically, that high concentrations of caffeine, administered to rats over several weeks in the water supply, induced paw and tail gnawing.

Purine nucleotide biosynthesis is regulated by three biochemical mechanisms. The first of these and almost certainly the most important is negative feedback control of glutamine-PRPP amidotransferase. This enzyme, which catalyses the first committed step in the biosynthetic sequence (Fig. 6.5) i.e. the amidation of PRPP by glutamine, has two separate allosteric sites, one specific for AMP, ADP and ATP, and the other for GMP, GDP and GTP. All three adenine nucleotides and all three guanine nucleotides inhibit, and it has been shown that when one allosteric site is saturated, further inhibition can still be obtained by adding a nucleotide of the other group. This is an example of concerted feedback control. The second control mechanism operates towards the end of the biosynthetic sequence at the point where it branches and IMP is converted into either adenine nucleotides or guanine nucleotides. As can be seen from Fig. 6.6, formation of the adenine nucleotides requires GTP and formation of the guanine nucleotides requires ATP. A positive feedback is thus achieved with a product of one branch of IMP metabolism accelerating the other. This has the effect of producing a balance in the output of adenine and guanine nucleotides. A third regulatory process is due to the inhibitory effect of AMP on the production of adenylosuccinate from IMP, and the inhibition by GMP of the oxidation of IMP to XMP (Fig. 6.6). This operates, like the major control on glutamine-PRPP amidotransferase, as a negative feedback on purine nucleotide synthesis. Additionally, when the overall energy status of the tissue is low (see 'energy-charge' concept, Section 6.6), the accumulating nucleoside diphosphates ADP and GDP further restrict purine nucleotide production. This is because ADP and GDP inhibit ribose-5-phosphate pyrophosphokinase, the enzyme producing PRPP from ribose 5-phosphate and ATP.

In those genera of higher plants that produce methylated xanthines, e.g. *Coffea* and *Camellia, de novo* purine biosynthesis is essentially similar to that of other plants and the IMP produced can therefore be aminated to AMP or oxidized to xanthosine 5′-monophosphate (XMP) and then aminated to GMP (Fig. 6.6). In tea (*Camellia sinensis*) and coffee (*Coffea arabica*) XMP has an additional fate. It can be dephosphorylated to xanthosine which is then either hydrolysed to xanthine for entry, *via* xanthine oxidase, into the pathway of purine catabolism (Fig. 6.7) or, methylated enzymically with *S*-adenosylmethionine (SAM) as the methyl donor, to yield 7-methylxanthosine. It is from this purine riboside that the other methylated xanthines then arise. The process involves enzymic hydrolysis to xanthine, followed by the series of methylations shown in Fig. 6.8 with SAM as the methyl donor. The biogenesis of methylated xanthines has been reviewed by Suzuki *et al.* (1992).

Until recently, an intriguing problem in purine biosynthesis has been the origin of the unsubstituted purine ring of the antibiotic nucleoside nebularine (12). This compound, first isolated from the fungus *Lepista (Clitocybe) nebularis*, has the structure purine 9-β-D-ribofuranoside. Although Fischer, in 1907, had predicted the natural occurrence of purine, this is the only known example. The particular biosynthetic problem posed by nebularine is that all naturally occurring purines arise from IMP, a 6-hydroxypurine derivative, and in consequence have either a 6-hydroxy substituent or a 6-amino substituent, and whereas such an amination is biochemically facile, complete removal of the 6-substituent, as in nebularine, is unique. The process has now been shown by Brown & Konuk (1995) to involve a novel enzymic mechanism which reductively deaminates adenosine, releasing the amino group as hydroxylamine (Fig. 6.9). During the course of this work, both nebularine 5′-phosphate, the corresponding nucleotide, and purine, the free base, were also shown to be natural products.

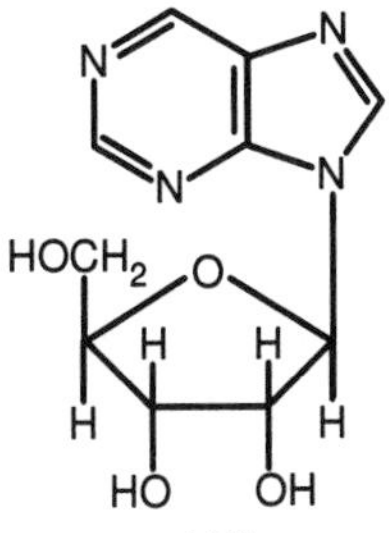

(12)

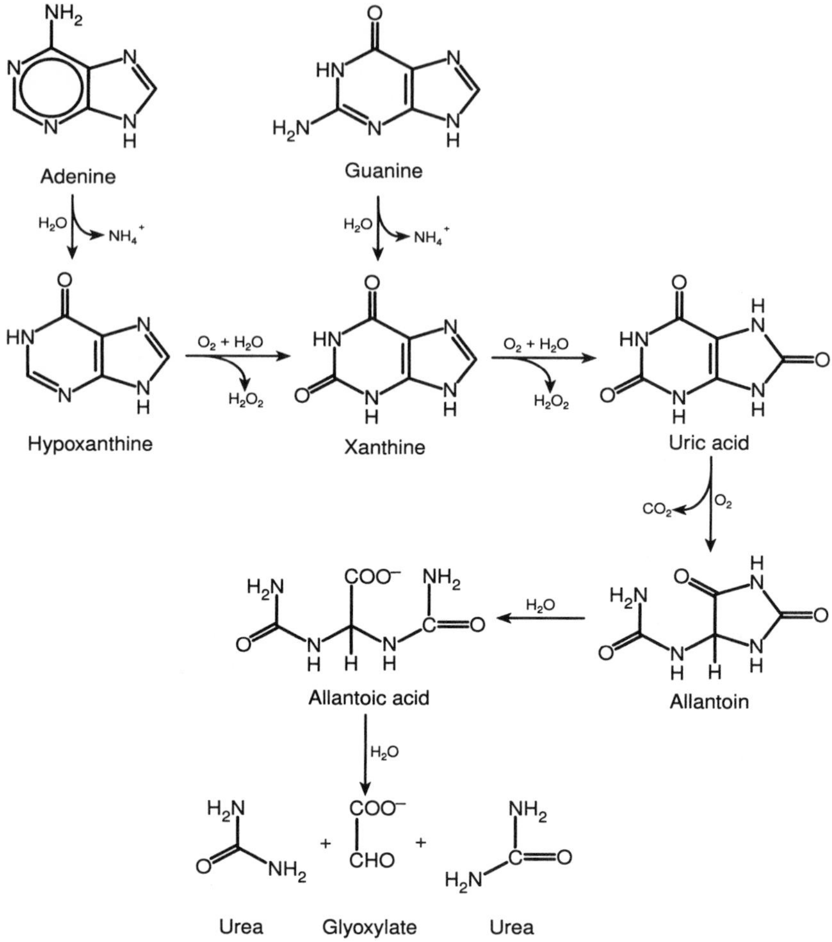

Fig. 6.7 Pathway of purine catabolism. The extent to which purine molecules are degraded in animals depends upon the type of organism concerned. Primates, including man, primarily excrete uric acid as the end-product of the catabolic process. Other mammals, turtles and molluscs excrete allantoin. Some fish excrete allantoic acid, most fish and amphibia excrete urea, and some invertebrates excrete ammonia as their main product of purine catabolism.

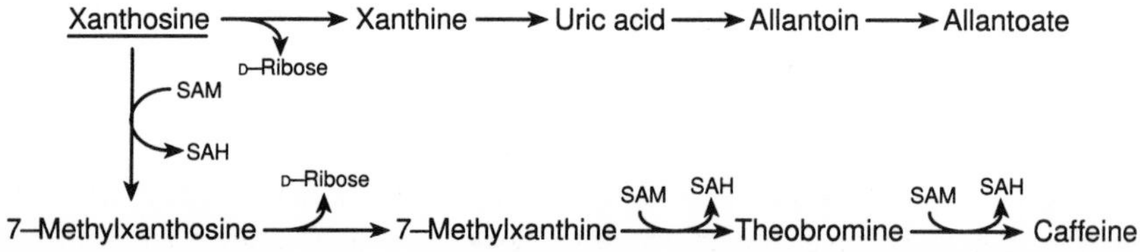

Fig. 6.8 Production of methylated xanthines by *Camellia sinensis* (tea) and *Coffea arabica* (coffee). SAM is the biological methylating agent S-adenosylmethionine which, after donating its methyl group, becomes S-adenosylhomocysteine (SAH).

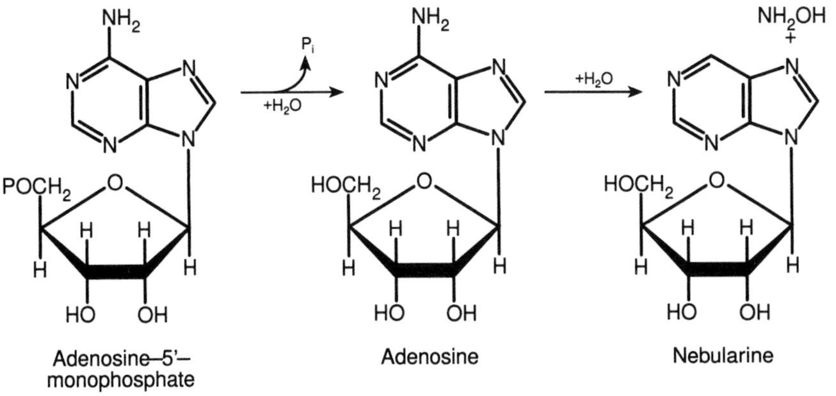

Fig. 6.9 Biosynthesis of purine riboside (nebularine). Nebularine is unusual in that it is a naturally occurring purine derivative without a substituent at C-6. It is formed from adenosine with loss of the 6-amino group as hydroxylamine.

6.4 PURINE CATABOLISM

Redundant nucleic acids and those of dietary origin are degraded by enzymic hydrolysis to a mixture of mainly 3′-nucleotides with some 5′-nucleotides; further hydrolysis dephosphorylates nucleotides to the corresponding nucleosides. In mammalian tissues at this catabolic stage, adenosine arising from AMP is deaminated enzymically to inosine. Guanosine and inosine are then hydrolysed to their corresponding bases guanine and hypoxanthine, and these are either recycled by the 'salvage pathways' (Section 6.3) or further catabolized according to need. In some organisms, adenosine is first hydrolysed to adenine and then deaminated to hypoxanthine. The pathway by which free purine bases are catabolized is outlined in Fig. 6.7. Xanthine oxidase, the key enzyme in this scheme, is a complex flavoprotein containing an atom of molybdenum and four iron-sulphur centres per molecule. It catalyses the oxidation of hypoxanthine to xanthine, and of xanthine to uric acid. Molecular oxygen serves as the electron acceptor for both steps. Uric acid from this oxidative sequence is ring-opened by the enzyme urate oxidase to yield the imidazole derivative allantoin. Under alkaline but non-enzymic conditions, urate rapidly undergoes a similar reaction in the presence of traces of Cu^{2+} ions and it is of interest that urate oxidase is a copper-protein. This is one of a number of examples of metal-catalysed reactions in Nature that evolution appears to have improved upon through development of a specific and highly efficient metalloprotein catalyst. Next, the imidazole ring of allantoin is hydrolytically opened in a reaction catalysed by allantoinase. Finally, the product allantoic acid is cleaved hydrolytically by allantoicase to yield urea and ammonia. In some organisms, e.g. marine invertebrates and

Table 6.3 The purine degradation products excreted by various types of organisms

Guanine	Spiders
Uric acid	Primates, birds, reptiles, and insects
Allantoin	Other vertebrates, some higher plants
Allantoic acid	Some teleost fishes, other plants
Urea	Amphibians, most fishes, and some molluscs
Ammonia	Marine invertebrates, and crustacea

crustacea, the urea is further hydrolysed by urease with the release of ammonia. The purine catabolic scheme shown in Fig. 6.7 is, in fact, a composite diagram and few organisms degrade purines to urea or ammonia. Most stop at some intermediate stage and excrete the product. In animals, the determining factors appears to be solubility of the product and availability of water in their immediate environment, either now or at some earlier point in the evolutionary time. In higher plants incomplete purine degradation resulting in formation of allantoin or allantoate is commonly seen. The comparative biochemistry of purine degradation is summarized in Table 6.3.

6.5 DEFECTS IN PURINE METABOLISM

The disease 'gout', which occurs in about 0.3% of the population, is due to malfunctioning of the metabolism of purines, and is characterized biochemically by an excessive accumulation of uric acid and its salts. Urates are of low solubility and their accumulation in plasma in concentrations greater than 0.5 mmol l^{-1} causes crystallization to occur, especially in the joints. Normally, joints are well lubricated and the granular deposition of urates, somewhat like having grit in the eye, causes pain and inflammation. In chronic cases, this can lead to a severe arthritic degeneration.

Gout usually results from overproduction of purine nucleotides due to one or more of the following biochemical aberrations: (a) elevation of the activity of 5-phosphoribosyl-1-pyrophosphate (PRPP) synthetase (the first enzyme in the purine biosynthetic sequence; Fig. 6.5), (b) loss of sensitivity of PRPP amidotransferase (the second enzyme; Fig. 6.4) to feedback inhibition by purine nucleotides, and (c) deficiency of the 'salvage' enzyme hypoxanthine-guanine phosphoribosyltransferase. Why this latter event should contribute to the aetiology of gout is not obvious but it may be due to the resulting decrease in use of PRPP and its consequent increased availability for purine biosynthesis.

Some twenty years ago, studies of the inherited condition 'severe combined immunodeficiency syndrome' led to the discovery that this also results from a defect in purine metabolism. Patients with this syndrome have defective immune systems which render them susceptible, often fatally so, to all kinds of infections. The biochemical defect has been identified as a genetic deficiency of adenosine deaminase, the enzyme that converts adenosine to inosine, and 2′-deoxyadenosine to 2′-deoxyinosine. In patients with the condition, the B and T lymphocytes which would normally proliferate in response to antigenic challenge cannot do so because the accumulating 2′-deoxyadenosine from DNA degradation is rapidly phosphorylated to 2′-deoxyATP, a potent inhibitor of deoxyribonucleotide synthesis and hence of DNA replication. Since 2′-deoxyATP kills white cells even when they are not proliferating, additional effects of this nucleotide may also be contributing to the problem.

Another, less severe, immunodeficiency syndrome described in recent years has also been shown to be due to a defect in purine metabolism. In this condition, the enzyme purine nucleoside phosphorylase is missing or defective. It appears to affect the T-cells specifically and causes them to accumulate 2′-deoxyGTP, another inhibitor of DNA replication. The B-cells remain unaffected.

6.6 BIOCHEMICAL FUNCTIONS OF PURINES

Like the pyrimidines, in addition to their role as nucleic acid constituents, the purines are key compounds in metabolism. They function, primarily in their free nucleotide form, in a variety of metabolic, cell-signalling, and regulatory processes. In particular, they are essential components of the energy transduction systems that drive all endergonic biochemical reactions. The central compound in bioenergetics is adenosine 5′-triphosphate (ATP) (13), an adenine nucleotide with a

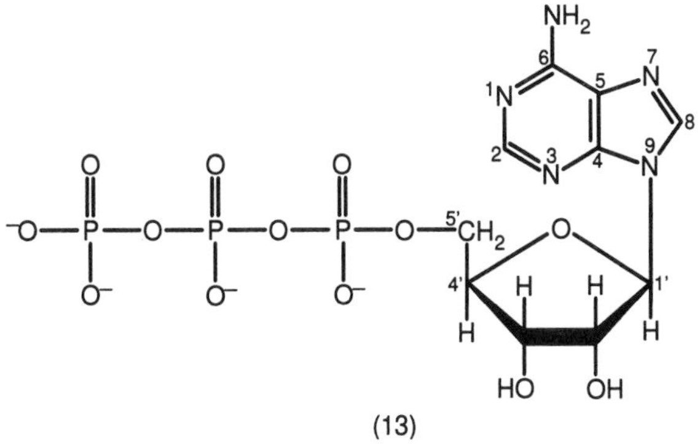

(13)

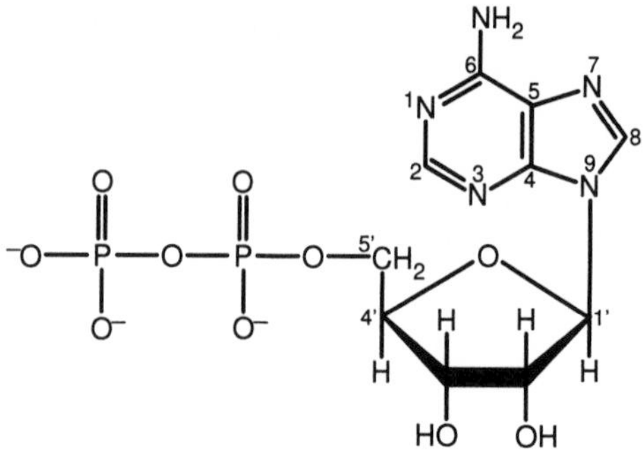

(14)

large free energy of hydrolysis ($\Delta G^{o\prime} = -8$ kcal mol^{-1}; pH 7) that can be harnessed by biological systems to drive endergonic reactions. The process involves transfer of the terminal phosphate group of ATP to a suitable acceptor, raising its chemical potential energy, and leaving adenosine 5′-diphosphate (ADP) (14). A common and persistent fallacy in biology is that this involves direct hydrolysis of ATP but such a process would result in a wasteful dissipation of energy as heat. Instead, the biochemical mechanism involves thermodynamic coupling of two reactions. This is exemplified by ester synthesis, represented chemically as:

(i) RCOOH + R′OH → RCOOR′ + H_2O, $\Delta G^{o\prime} = +8$ kJ mol^{-1}

In a biological system this could be effected by the coupling of reactions (ii) and (iii) the summation of which is shown in equation (iv); in these reactions *P* represents an orthophosphate group:

(ii) RCOOH + ATP → RCOO*P* + ADP, $\Delta G^{o\prime} = -12.5$ kJ mol^{-1}

(iii) RCOOP + R′OH → RCOOR′ + *P*.OH, $\Delta G^{o\prime} = -12.5$ kJ mol^{-1}

(iv) R.COOH + ATP + R′OH → RCOOR′ + ADP + *P*OH, $\Delta G^{o\prime} = -25$ kJ mol^{-1}

In effect, reactions (ii) and (iii) are linked thermodynamically through the intermediate formation of a high-energy phosphoanhydride (RCOO*P*) which is common to both reactions, and the overall process (iv) is exergonic. In a biological system, the reactions would be catalysed by one or more enzymes but it is important to remember that catalysts only affect velocity and not thermodynamic feasibility.

In effecting some endergonic reactions, cells make use of nucleoside triphosphates other than ATP. For example, GTP the guanosine analogue of ATP is specifically required in the reaction catalysed by phosphoenolpyruvate carboxykinase:

$$\text{Oxaloacetate} + \text{GTP} \rightleftharpoons \text{phosphoenolpyruvate} + CO_2 + \text{GDP}$$

also in the formation of succinyl CoA from succinate by succinyl CoA synthase:

$$\text{Succinate} + \text{GTP} + \text{CoA.SH} \rightleftharpoons \text{succinyl-SCoA} + \text{GDP} + P_i$$

Nevertheless, these other nucleoside triphosphates are secondary to ATP and their production is ultimately at the expense of ATP being dephosphorylated to ADP, commonly through enzyme catalysed reactions of the type:

$$\text{ATP} + \text{GDP} \rightleftharpoons \text{ADP} + \text{GTP}$$

The equilibrium constant for this reaction is essentially unity but the relatively much higher concentration of ATP favours synthesis of the other nucleoside triphosphate.

ATP is present in living cells in limited concentrations (*ca.* 2–5 mM) and has to be continually recycled. Rephosphorylation of ADP to ATP is the main function of the cellular respiratory processes. Some of this phosphorylation is achieved during anaerobic respiration in the cytosol. Inorganic phosphate is used enzymically in that process to convert glyceraldehyde 3-phosphate into glyceric acid 1,3-bisphosphate (reaction v). This high energy compound (i.e. possessing a high free energy of hydrolysis) is used, in turn, to phosphorylate ADP to ATP (reaction vi):

(v) glyceraldehyde 3-P + NAD^+ + P_i → glyceric acid 1,3-bis$P$ + NADH
(vi) glyceric acid 1,3-bis$P$ + ADP → glyceric acid 3-P + ATP

By far the biggest sources of ATP in biological systems are however the membrane-bound enzyme complexes associated with aerobic respiration and, in green plants, with photosynthesis. These energy-transducing membranes are the plasma membranes of prokaryotic cells such as bacteria and cyanobacteria, the inner mitochondrial membranes of eukaryotes (animals and higher plants), and the photosynthetic thylakoid membranes of chloroplasts in green plants and photosynthetic microorganisms. The complex mechanisms involved have been the subject of intensive biochemical research for over half a century but the unifying principle which is the key to all of them was elucidated in 1961 by Peter Mitchell. The 'chemiosmotic hypothesis', as it is called, has probably been more rigorously tested than any other biochemical theory and it can be said to have gained ultimate acceptance in 1978 when Mitchell was awarded the Nobel Prize in Chemistry for his work. In outline, the mechanism involves the establishment of a proton gradient across the intact inner membrane of the organelle and its maintenance by the electron transport processes of respiration or photosynthesis which occur inside the membrane. The high concentration of protons outside the membrane causes their translocation back through special proton channels associated with the enzyme ATPase. The resulting proton motive force is required for the dissociation, from the ATPase, of ATP formed spontaneously at the active site. In effect the ATPase reaction is driven to the left:

$$ATP + H_2O = ADP + P.OH$$

For many years, it had puzzled biochemists as to why the energy-generating organelles should have so much ATP-hydrolysing activity when this would degrade the ATP synthesized; a classical case of looking at something the wrong way around. The high ATPase activity is essentially an artefact of disrupted or damaged membranes in contrast to the ATP synthase activity of intact membranes. For a more detailed account of Mitchell's work and the chemiosmotic hypothesis, the reader is referred to Nicholls and Ferguson (1992).

Because of its dominant central role in energy transduction by all living organisms, ATP has often been described as the energy currency of the cell. It is the immediate energy source for all aspects of the routine metabolism of cells and also for the more specialized energy-transduction processes of muscular contraction, luminescence (e.g. in fireflies and glow-worms), and the generation of electricity by the electric organs of certain fishes (e.g. *Torpedo, Raia, Gymnotos* and *Malapterurus*). The ubiquity of ATP inevitably poses the question as to what it is about the structure and properties of this molecule that make it so pre-eminently suitable for its role in energy transduction. Consideration of this question leads to the conclusion that it is not just the relatively large negative free energy change associated with the hydrolysis of ATP to ADP (–31 kJ mol^{-1}). Important though this is, it is not unique amongst biological molecules. The glycolytic intermediate phosphoenolpyruvate, for example, has a $-\Delta G^{o\prime}$ of hydrolysis twice that of ATP at –62 kJ mol^{-1}. The large $-\Delta G^{o\prime}$ is however of little consequence if the compound cannot transfer its phosphate, and hence its energy potential, to a variety of biochemical acceptors. With phosphoenolpyruvate, this transferability is largely restricted to ADP during glycolysis, and to certain sugar acceptors during transport of the latter through membranes by the phosphotransferase system. In biological energy transduction, coupling between exergonic and endergonic reactions is accomplished by specific enzymes. The utility of one high energy compound versus another is therefore also determined by its ability to fit the appropriate part of the enzyme topology. This is not only a matter of size and shape but also of charge distribution.

During coupling reactions involving ATP, it is the terminal phosphate group that is transferred, leaving the diphosphate ADP. In a few instances, two phosphate groups are transferred from ATP, *en bloc*, as pyrophosphate, leaving behind the monophosphate, AMP. An example of this is seen in the first step in purine biosynthesis (Fig. 6.5) in which

5-phosphoribosyl 1-pyrophosphate (PRPP) is formed from ribose 5-phosphate and ATP. As the high energy potential of ATP is associated with the phosphoanhydride structure of the two terminal phosphate groups, both ATP and ADP are, in effect, high energy compounds. Loss of both of these phosphoanhydride groups leaves AMP which is a low energy phosphate ester.

Because of the central role of the adenine nucleotides in bioenergetics, the relative concentration of these compounds in a given tissue has been used as an indicator of the metabolic status of that tissue. In simple terms, relatively high concentrations of ATP and ADP, the high energy compounds, in comparison with AMP, the low energy compound, indicates that the tissue has a high potential for undertaking useful biological work, i.e. to carry out its essential endergonic reactions. Atkinson has suggested that this approach can be put onto a more quantitative basis by using the 'energy-charge' of the adenylate pool as a regulatory parameter. Using the analogy of a lead accumulator cell, in which electrical charge is a function of the relative amounts of the three oxidation states of lead,

$$2PbSO_4 + 2H_2O = Pb + PbO_2 + 2H_2SO_4$$

Atkinson (1977) has suggested that the appropriate charge function to consider with the adenylate system in cells is the number of anhydride-bound phosphates per adenosine moiety. This would vary from 0 at complete discharge (only AMP present) to 2 at full charge (only ATP present). Dividing by 2 gives a parameter ranging from 0 to 1 and this is Atkinson's 'energy charge'. It is in practical terms:

$$\frac{[ATP] + 0.5[ADP]}{[ATP] + [ADP] + [AMP]}$$

Since most tissues contain the enzyme adenylate kinase which catalyses the reaction

$$AMP + ATP = 2ADP$$

only AMP and ATP are usually present in significant concentrations so that the equation reduces to:

$$Energy\ charge = \frac{[ATP]}{[ATP] + [AMP]}$$

The relationship of energy charge to metabolic regulation is best seen from Fig. 6.10 which shows a generalized response of the activity of regulatory enzymes to changes in this parameter. The cross-over point at an energy charge of approximately 0.85 represents a stable meta-

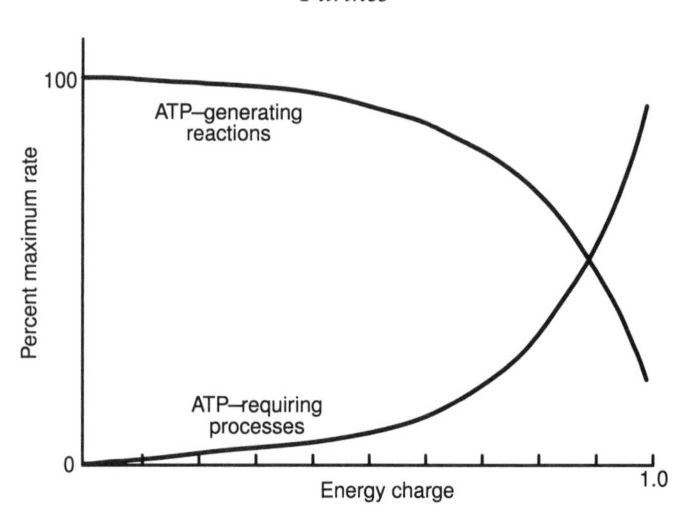

Fig. 6.10 Relationship of energy charge to the rate of key enzymes in metabolism. The mechanism of this effect is that the key enzymes are controlled allosterically by adenine nucleotide concentrations. ATP-generating reactions are turned off as ATP accumulates whilst simultaneously those requiring ATP, e.g. synthetase reactions, are switched on. The cross-over point is at an energy charge of approximately 0.85 and represents a stable metabolic state.

bolic state. Any decrease in energy charge increases the rates of the ATP-generating sequences and simultaneously decreases those of ATP-consuming processes. This does not apply to all the reactions of ATP-generating or ATP-consuming sequences but only to the rate-controlled steps. The mechanism of this effect is simply that the key enzymes are allosteric and their catalytic activity is regulated by adenine nucleotides. An example from an ATP-generating process is phosphofructokinase, a rate-controlling enzyme in glycolysis. This enzyme is activated allosterically by AMP and ADP and inhibited by ATP. Accumulation of AMP or ADP thus switches on glycolysis to generate ATP, and an abundance of ATP switches it off. Another example is pyruvate dehydrogenase. This key enzyme in the ATP-generating tricarboxylic acid cycle is similarly allosterically inhibited by an abundance of ATP. An example on the ATP-utilizing side of metabolism, is phosphoribosyl pyrophosphate (PRPP) synthetase, a key enzyme in *de novo* purine biosynthesis. AMP and ADP inhibit this enzyme allosterically and thus switch it off under conditions of low energy charge, so conserving ATP.

A function of adenine nucleotides which is at once obvious and yet obscure, is their role in coenzyme structure. Many coenzymes and enzymic prosthetic groups have a molecule of AMP incorporated into

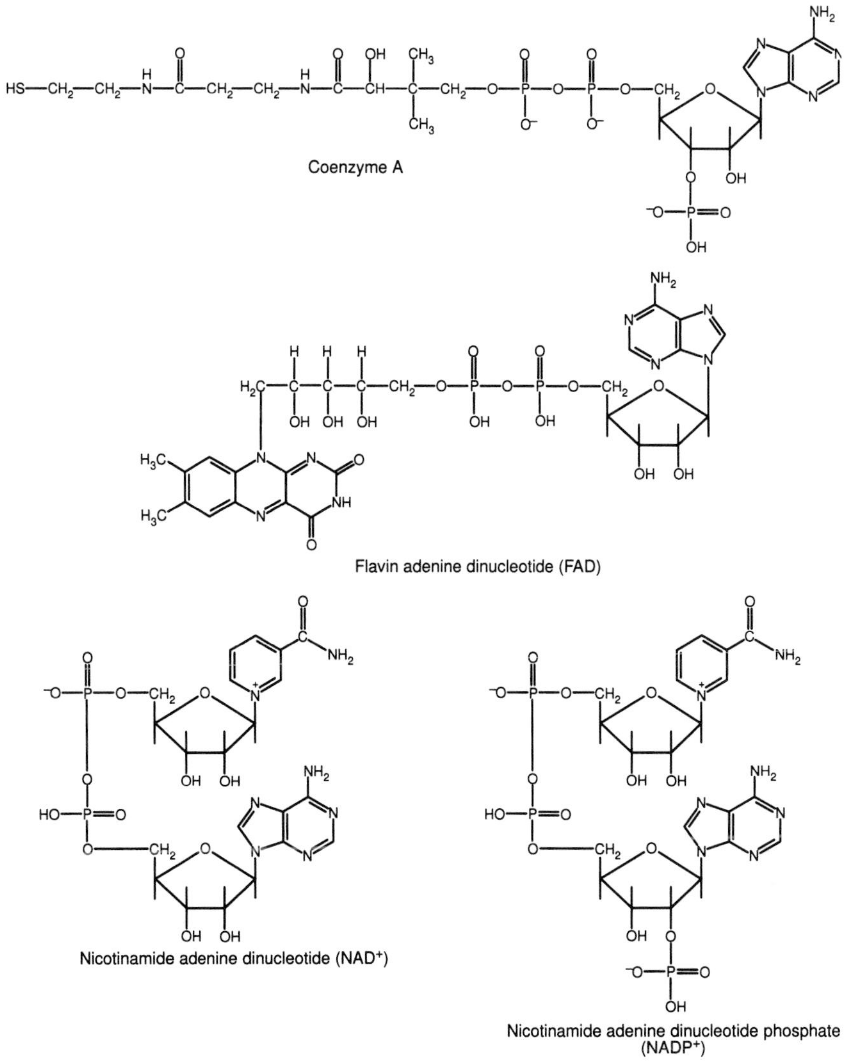

Coenzyme A

Flavin adenine dinucleotide (FAD)

Nicotinamide adenine dinucleotide (NAD^+)

Nicotinamide adenine dinucleotide phosphate ($NADP^+$)

Fig. 6.11 Coenzymes and enzyme prosthetic groups having a molecule of adenosine 5′-monophosphate (AMP) built into their structure.

their structure. Examples of this (Fig. 6.11) are the pyridine nucleotides NAD^+ and $NADP^+$, and the flavin nucleotide FAD, all of which are essential redox components of many enzymes; and coenzyme A, the coenzyme of biochemical acylations. There is also an adenosine residue within the molecular structure of *S*-adenosylmethionine (15), the coenzyme of sulphur metabolism; and in 5′-deoxyadenosylcobalamin, the coenzymic form of vitamin B_{12} (16). With none of these coenzymes and prosthetic groups is there however any indication that the adenine

(15)

(16)

nucleoside or nucleotide component plays an active role in the reaction mechanisms being catalysed. It has to be concluded therefore that the role of the purine unit is related to the specific docking of the coenzyme at the correct location on the apoenzyme surface. Similar considerations would explain the frequent occurrence of adenine and other purine nucleotide derivatives as allosteric regulators which bind to enzyme proteins at specific but non-catalytic sites altering conformation and hence catalytic efficacy.

Cell-signalling is another area of major biochemical importance dominated by purine nucleotides. The discovery by Sutherland and his associates in 1957 of a dialysable, heat-stable factor that mediated the effect of adrenalin on liver phosphorylase opened this field of continuing high research activity by, in effect, showing for the first time how a hormone works at a molecular level. Whereas much was already known of the physiological effects of hormones, how these messenger molecules are transduced into biochemical activity was obscure until Sutherland's Nobel Prize winning discovery. The dialysable, heat-stable cofactor was subsequently isolated by Sutherland and identified as adenosine 3′,5′-cyclic monophosphate (17). How this identification was made, recounted by Sutherland (1971), is an interesting example of scientific coincidence.

> '... We asked Dr Leon Heppel for an enzyme that might help us characterize a substance that we had isolated from animal tissue preparations. He kindly supplied us with this enzyme; the enzyme did not hydrolyse our compound, a result we had not expected. We thanked Dr Heppel and in our letter stated: "However, it did not attack our nucleotide (or polynucleotide) and, therefore, did not support our tentative idea that the compound may be a double diester, i.e. a dinucleotide...". Some weeks later Dr Heppel called us stating that Dr David Lipkin had written ... to ask for the same enzyme to help characterize an unknown nucleotide that his group had isolated from a barium hydroxide digest of ATP. Dr Heppel informed us and Dr Lipkin that the tentative structures proposed by both groups were identical. Very soon the groups exchanged samples and found that they were indeed identical. ... Dr Lipkin ... completing the determination of structure and molecular weight (Lipkin *et al.*, 1959); thus two very amateur chemists – Dr Rall and I – were freed so that we could continue our studies of the formation of this chemical agent in animal tissues.'

Cyclic AMP is now known to mediate the effect of other hormones (Table 6.4) and in a cell-signalling context is often described as a 'second messenger', the hormone being the primary messenger. The mechanism, outlined in Fig. 6.12, involves binding of the hormone at a receptor in the target-cell membrane and this, in turn, causes the α-

Table 6.4 Hormones using cyclic AMP as a second messenger

Hormone
Adrenalin (epinephrine)
Calcitonin
Choriogonadotrophin
Corticotropin (ACTH)
Follicle-stimulating hormone
Glucagon
Gonadotropin-releasing hormone
Growth hormone-releasing hormone
Lipotropin
Luteinizing hormone
Melanocyte-stimulating hormone
Parathyroid hormone
Secretin
Thyroid-stimulating hormone
Vasopressin

subunit of a transduction protein (G-protein) to be activated by GTP. The newly activated α-subunit, with GTP attached, then binds to, and activates, the enzyme adenylyl cyclase, also located within the membrane. Adenylyl cyclase catalyses the formation of cyclic AMP:

$$\text{ATP} \rightarrow \text{cyclic AMP} + \text{PP}_i$$

The cyclic nucleotide activates protein kinases which catalyse phosphorylation of inactive enzymes, so activating them. The sequence of events from cyclic AMP to the physiological response is, in effect, a cascade in which one catalyst activates a much greater amount of another, which in turn activates even more molecules of another, i.e. each stage amplifies substantially the effect of the previous one until the full physiological response is manifested.

Central to this scheme (Fig. 6.12) are the G-proteins, which specifically bind guanosine 5′-diphosphate and -triphosphate. The GTP-bound form is the active factor in switching-on adenylyl cyclase but this G-protein also possesses its own GTPase to hydrolyse the triphosphate back to the diphosphate, so preventing a perpetual activation of the cyclase. Any efficiently functioning signalling system must have such signal-dampening mechanisms to prevent continuous response; another example is seen with cyclic AMP itself and which is continually removed from the system by a cyclic nucleotide phosphodiesterase catalysing the reaction:

$$\text{cyclic AMP} + \text{H}_2\text{O} \rightarrow \text{AMP}$$

In microorganisms, cyclic AMP is involved in other signalling and regulatory processes. The role of cyclic AMP in the life cycle of the slime mould *Dictyostelium discoideum* is particularly interesting. In conditions of plenty with a ready supply of nutrients, *Dictyostelium* exists in the

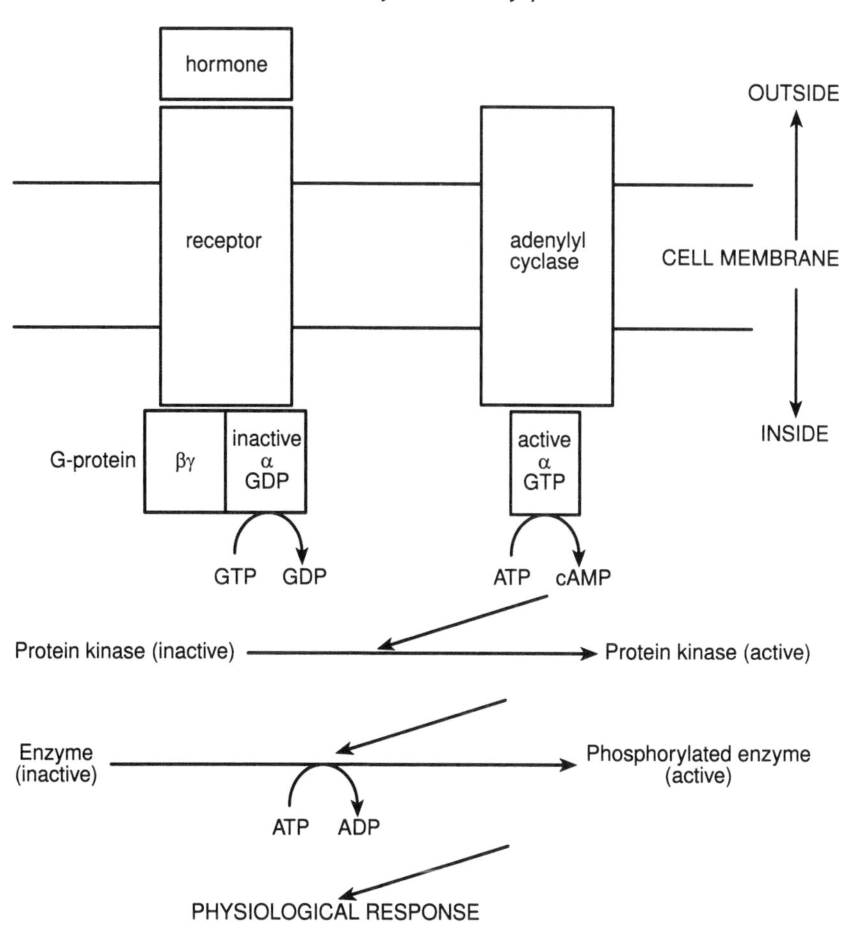

Fig. 6.12 Cyclic AMP second messenger system. A hormone operating through this mechanism, binds at a receptor causing the α-subunit of a G-protein to be activated by GTP. The activated subunit, in turn, activates adenylyl cyclase which catalyses formation of cyclic AMP from ATP. The second messenger, cyclic AMP, then causes phosphorylation of inactive enzymes into the phosphorylated, active form. The physiological response follows. The signal is 'switched-off' as GTP is hydrolysed and adenylyl cyclase again becomes inactive.

form of independent, free-living amoebae. When the supply of nutrients begins to dwindle, cyclic AMP is secreted by the cells and serves as a chemoattractant so that swarming occurs and the amoebae aggregate into a slug-like mass which then structures itself into a fruiting body. In this function, cyclic AMP could be regarded as a primary messenger.

Similarly, in bacteria cyclic AMP is involved in a system monitoring levels of nutrients. When the concentration of glucose falls, cyclic AMP levels rise and activate the lactose (*lac*) operon by virtue of its interac-

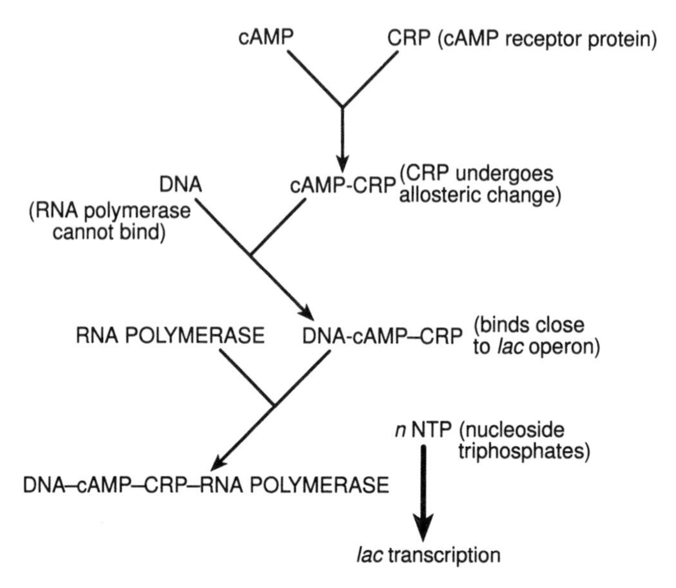

Fig. 6.13 Regulatory role of cyclic AMP during transcription in bacterial protein synthesis. This example shows the transcription of the *lac* operon in *E. coli*. The cyclic nucleoside complexes with the receptor protein (CRP) which, in turn, binds with the region of DNA close to the *lac* promoter. This facilitates access of RNA polymerase and so initiates transcription.

tion with a cyclic AMP receptor protein (CRP). The interaction causes a conformational change to occur in the CRP, greatly increasing its affinity for certain sites on DNA, including a site in the *lac* operon which is next to the RNA polymerase binding site. This facilitates transcription of the *lac* operon, i.e. the enzyme β-galactosidase is synthesized, enabling β-galactosides, such as lactose, to be hydrolysed thus making available a wider range of carbohydrates as nutrients. The biochemical events in the process are outlined in Fig. 6.13.

Guanosine 3′,5′-cyclic monophosphate (18), also known as cyclic GMP, is the guanine analogue of cyclic AMP (17), and plays a major signal-transducing role in the visual process (Fig. 6.14). The outermost segment of each retinal rod consists of a specialized photosensitive system constructed of lamellated membranes. These membranes contain the photoreceptor pigment rhodopsin, the non-protein component of which is 11-*cis*-retinal, a derivative of vitamin A. Light triggers the conversion of 11-*cis*-retinal into the all-*trans* form and in the dark the process reverses. This *cis-trans* isomerization of retinal and the consequent conformational changes in the protein moiety of rhodopsin are

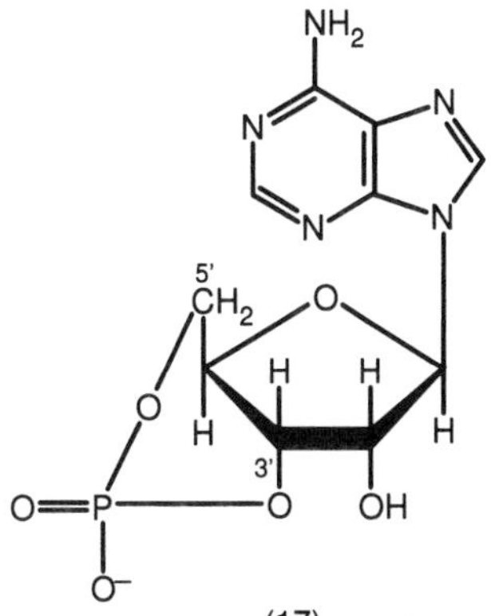

(17)

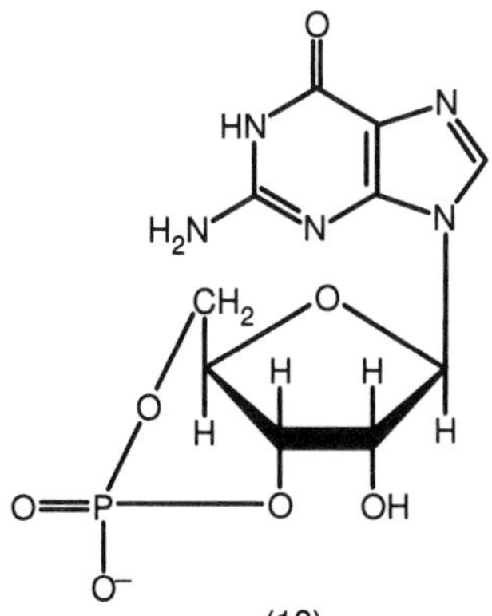

(18)

the primary events in visual excitation. Transduction of these primary events into a nerve impulse is mediated by cyclic GMP. What happens is that the photoexcited rhodopsin triggers a series of reactions resulting in hydrolysis of cyclic GMP. As, however, the concentration of cyclic GMP controls cation-specific channels in the boundary membrane of the outer rod segment, the effect of light is to halt the ingress of Na^+ ions so producing a rapid hyperpolarization of the membrane. It has been calculated that, in this way, a single photon closes hundreds of the cation channels in a dark-adapted rod, causing a hyperpolarization of about 1 mV. This change in membrane potential is signalled, *via* the retinal neurons and the optic nerve, to the brain.

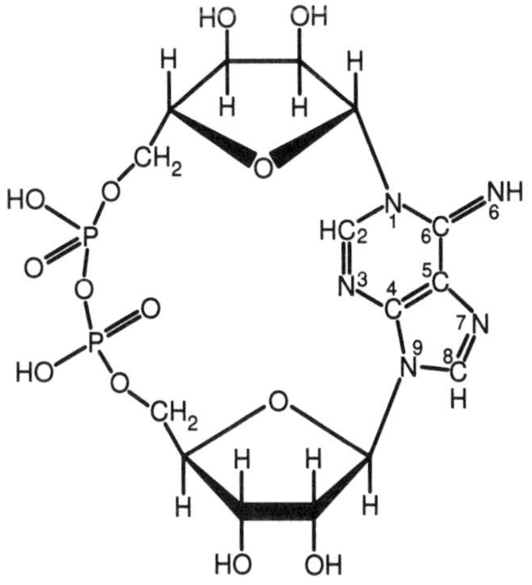

(19)

Recent research has revealed another signal-transducing nucleotide, cyclic ADP-ribose (19). This was discovered as a Ca^{2+} ion-mobilizing

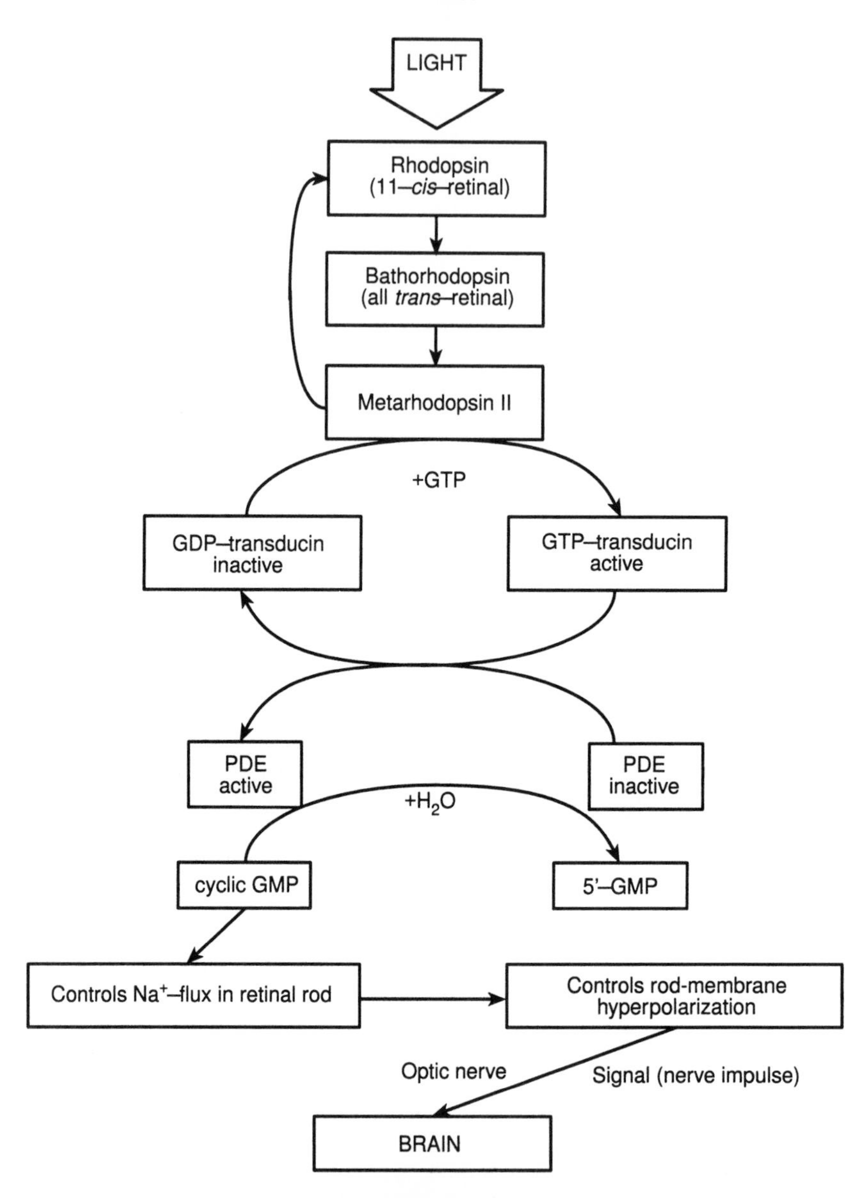

Fig. 6.14 The main biochemical processes in vision, showing the major role played by cyclic GMP. Light triggers the isomerization of 11-*cis* retinal to the all-*trans* form. This event brings about the activation of transducin by GTP, resulting in the activation of a cyclic nucleotide phosphodiesterase. In turn, cyclic GMP is hydrolysed and as the concentration of this nucleotide falls, the ingress of Na^+ ions into the boundary membrane of the outer rod segment is halted. The change in membrane potential is signalled *via* the retinal neurons and the optic nerve, to the brain.

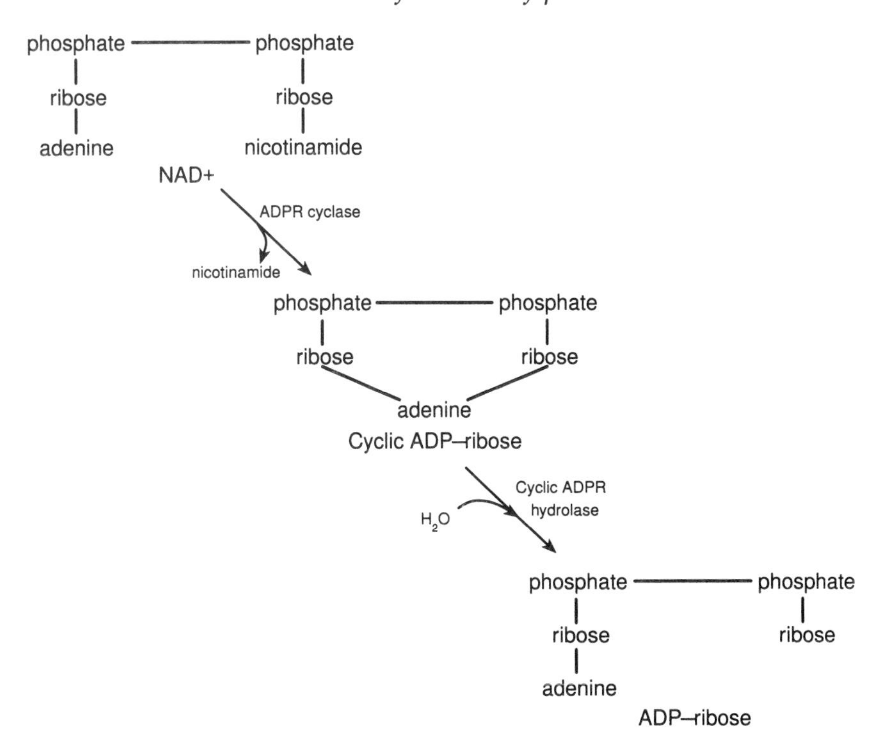

Fig. 6.15 Cyclic ADP-ribose signalling system. Cyclic ADP-ribose is synthesized from NAD^+ by a GTP-activated cyclase. Its production is accompanied by release of Ca^{2+} ions which activates other enzymes (e.g. protein kinases) causing biochemical changes to occur in the tissue concerned. The signalling compound cyclic ADP-ribose is removed, i.e. the signal is switched-off, by a specific hydrolase.

agent in sea urchin eggs but has since been found in various mammalian tissues and tissues from amphibia, birds and insects. It is formed from NAD^+ by the enzyme ADPR cyclase (Fig. 6.15) which is activated by cyclic GMP. Generation of cyclic ADP-ribose is followed by a release of Ca^{2+} ions which, in turn, activates other enzymes (e.g. protein kinases) causing biochemical changes to occur. This effect of Ca^{2+} ions can be direct, or indirect through the Ca^{2+}-binding protein calmodulin. Dampening of the signal in these cases is provided by the enzyme ADP-ribose hydrolase which hydrolyses the N^1-glycosidic bond of cyclic ADP-ribose to release ADP-ribose (Fig. 6.15).

In bacteria, metabolism has a 'fail-safe' control known as the 'stringent response'. When one or more amino acids are in short supply, a number of biosynthetic processes are shut down. This avoids wasting valuable resources on futile anabolism. The process is essentially a blocking of protein synthesis and hence of enzyme production. It is triggered by an accumulation of unusual guanine nucleotides, synthesized

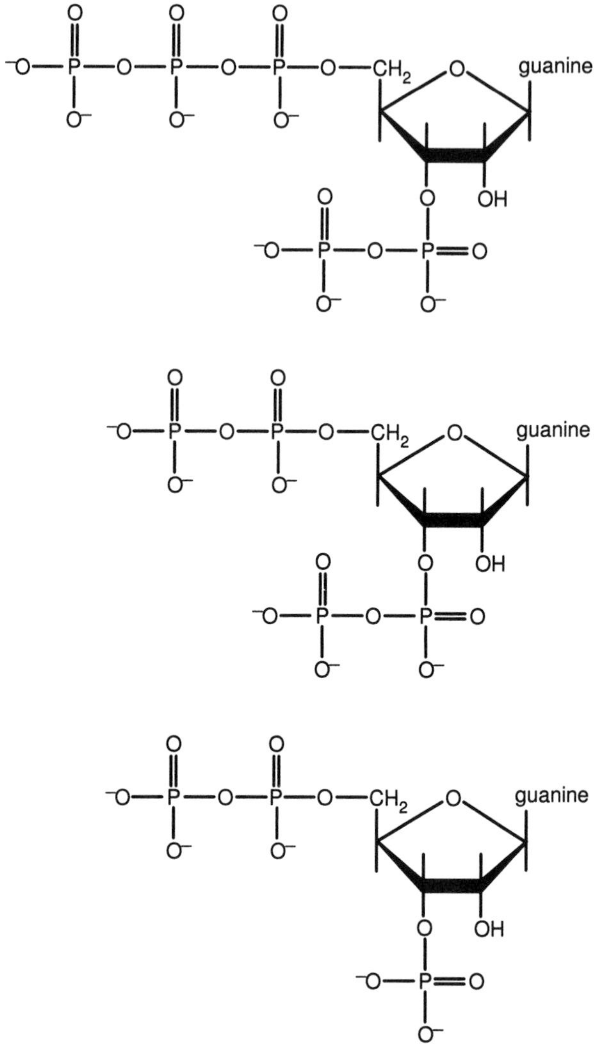

pppGpp

ppGpp

ppGp

Fig. 6.16 The guanine nucleotides controlling the 'stringent response' of bacterial cells. This is essentially a shut-down of protein synthesis in response to decreasing supplies of amino acids.

by ribosomes in the presence of ATP, GDP, mRNA and free tRNA (i.e. tRNA not carrying amino acids for protein synthesis). The compounds concerned (Fig. 6.16) are guanosine 5′-diphosphate-3′-diphosphate (ppGpp) also known as guanosine tetraphosphate, guanosine 5′-triphosphate-3′-diphosphate (pppGpp) and guanosine 5′-diphosphate-3′-monophosphate (ppGp).

A group of structurally related adenine nucleotides (Fig. 6.17) are

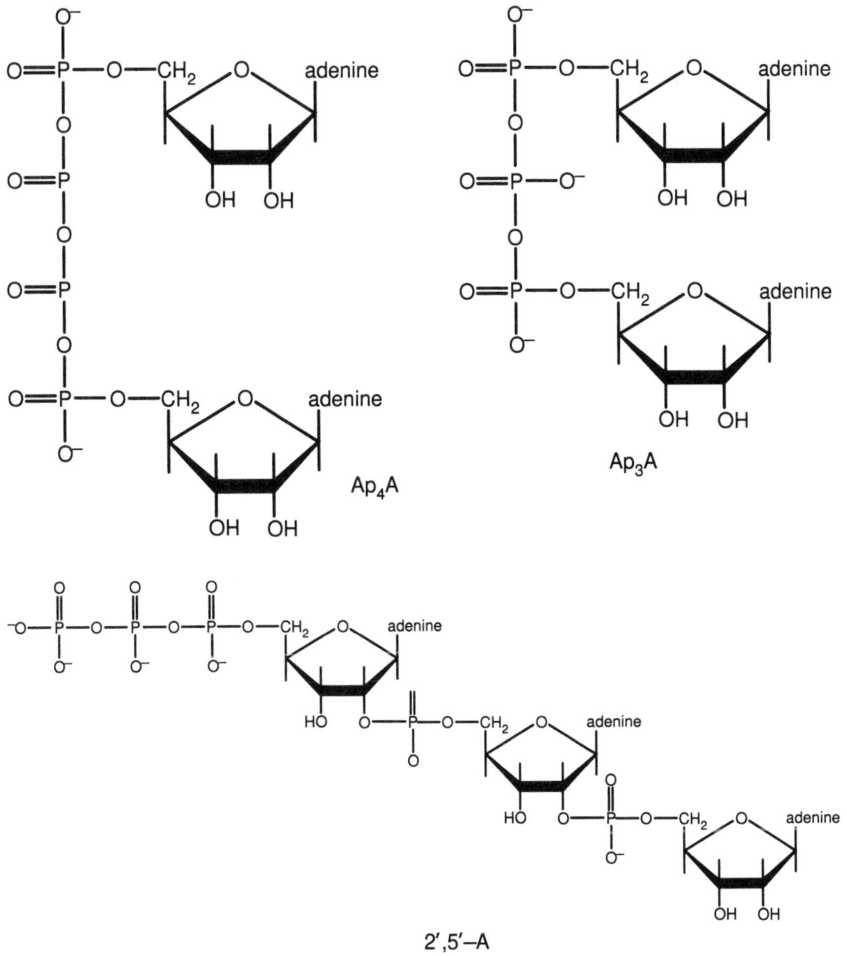

Fig. 6.17 A group of adenine oligonucleotides currently under investigation as cell-signalling compounds. The concentration of Ap_4A is 100-fold higher in S-phase cells than in cells arrested in the G_0/G_1 phase of the cell cycle. Ap_3A is a signalling compound that, at least in bacteria, is involved in enabling cells to respond rapidly to oxidative stress. 'Two-five A' (2′,5′-A) is involved in the mechanism by which interferon enables cells to resist attack by RNA viruses.

synthesized in all cells by aminoacyl-tRNA synthases in a reaction in which the AMP moiety of an aminoacyl-AMP is transferred to another adenine nucleotide. For example, when ATP is the acceptor, the compound formed is diadenosine tetraphosphate (Ap_4A). When ADP is the acceptor, Ap_3A is formed; if GTP is the acceptor, Ap_4G is formed. Evidence is accumulating that these nucleotides are involved in cell signalling and that some of them may be growth signals. The concentration of Ap_4A, for example, has been reported to be 100-fold higher in S-phase cells than in cells arrested in the G_0/G_1 phase of the cell cycle. It has been suggested that, at least in bacteria, Ap_3A is a signalling compound that enables cells to react rapidly to oxidative stress.

An adenine trinucleotide (pppA2′p5′A2′p5′A) known as 'two-five A' (Fig. 6.17) is produced as part of a mechanism by which interferon enables cells to resist attack by RNA viruses. It inhibits protein biosynthesis by activating an endonuclease that degrades cellular and viral RNA.

When cells sustain damage to their DNA, mechanisms are brought into play which repair that damage. However, if substantial damage occurs and in consequence a number of gaps appear in single-stranded regions, an enzyme called poly(ADP)-ribosyltransferase is activated. This catalyses the transfer of ADP-ribose residues from NAD^+ to the carboxyl groups of nuclear proteins and in turn causes a fatal depletion of cellular NAD^+. It has been suggested that this mechanism is an important part of 'programmed cell death' by which extensively damaged cells are destroyed, preventing them from becoming cancer cells. The extremely potent nature of both the cholera and the diphtheria toxins are due to their catalysis of similar ADP-ribosylations with NAD^+ as donor. In the case of diphtheria, the recipient is a modified histidine residue within the structure of an essential component of protein biosynthesis, known as elongation factor 2 (translocase); its ADP-ribosylation effectively blocks eukaryotic protein synthesis. With the cholera toxin, it is the activated G-protein of the cyclic GMP cell-signalling system that is the ADP-ribosylation target. The result is that the G-protein (Fig. 6.12) becomes permanently locked in the activated state causing persistent cyclic AMP activity. In turn, this continuously activates ion-pumps, leading to a large efflux of Na^+ ions and water into the gut and giving rise to the characteristic diarrhoea of cholera.

The nucleoside adenosine plays an important regulatory role in the central nervous system. It binds to membrane-bound cell receptors that regulate the formation of cyclic AMP by adenylyl cyclase. Through this mechanism, adenosine functions pharmacologically as a depressant and is antagonized at the receptors by methylxanthines, such as caffeine, which act as stimulants.

6.7 BIOLOGICAL AND PHARMACOLOGICAL ACTIVITY

In addition to the biochemical functions served by the diverse compounds described in the preceeding section (6.6), some naturally occurring purine derivatives exhibit pronounced biological and pharmacological activity. The methylated xanthines, in particular, fall into this category. Caffeine, theobromine and theophylline (Fig. 6.1) are all diuretics and have been used clinically for this purpose. Additionally, caffeine is used as a cardiac, respiratory, and psychotherapeutic stimulant, and has been used in the treatment of migraine. Theophylline is of value clinically in treating asthmatic bronchospasm, coronary insufficiency, hypertensive headache, biliary colic, Cheynes-Stokes respiration, congestive heart failure, and pruritus.

The cytokinins (Fig. 6.2) are a group of natural products having a pronounced stimulatory effect on plant cell division (Section 6.1). Most of them are N^6-substituted adenines; in the majority of cases the substituent is isoprenoid. The group is exemplified by zeatin (6-[4-hydroxy-3-methyl-*trans*-2-butenylamino]purine) which was the first naturally occurring cytokinin to be isolated (Fig. 6.2). Both the 9-β-D-ribofuranoside and its 5′-monophosphate, i.e. the corresponding nucleoside and nucleotide have also been found in plant extracts. Around 20 common, naturally occurring, cytokinins have been described although it seems unlikely that all of the are of physiological significance; a number may be metabolic products of physiologically functional compounds. They are generally present in plants at very low concentrations, typically in the ng per g fresh weight range and are formed both by enzymic isopentenylation of AMP with Δ^2-isopentenyldiphosphate as the donor, and by enzymic hydrolysis of isopentenylated mRNA.

In addition to their activity as cell division promoters in plant tissue cultures, two other biological activities of cytokinins are used as the basis for bioassays for these compounds. They are, (i) induction of betacyanin synthesis by tissues of *Amaranthus caudatus* in which the characteristic red-purple colour of the betacyanins facilitates spectrophotometric assay, and (ii) delaying of the senescence of detached wheat leaf segments. In this assay, pieces are cut from the leaves of wheat or of another monocotyledonous plant and floated on water for a day or two. They rapidly turn yellow as chlorophyll and protein are catabolized. Cytokinins, however, significantly delay these processes and their effects can be quantified by spectrophotometric measurements of chlorophyll concentration.

An unusual, naturally occurring analogue of guanine, 8-azaguanine (20), is produced by strains of the mould *Streptomyces albus*. This compound is the antifungal antibiotic Pathocidin which inhibits formation of GMP and is incorporated into the nucleic acids of various biological systems. 8-Azaguanine has found some clinical applications in cancer chemotherapy.

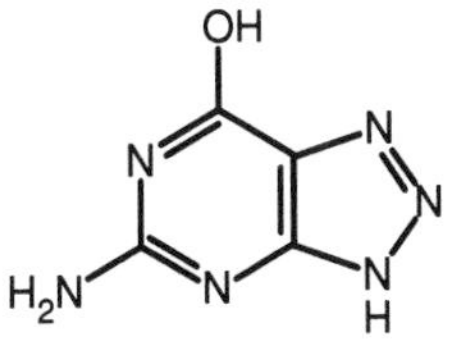

OH
N N
N
H_2N N N
H

(20)

6.8 SYNTHETIC PURINES AND PURINE ANALOGUES OF CLINICAL IMPORTANCE

One of the best known and extensively used purine analogues is allopurinol (4-hydroxypyrazolo[3,4-*d*]pyrimidine). This compound, used in the treatment of gout and related disorders of purine catabolism, is an analogue (21) of hypoxanthine (22) in which *N*-7 and *N*-9 of the latter are juxtaposed so that the imidazole ring of the purine becomes a pyrazole ring. Administered allopurinol is oxidized by xanthine oxidase, a key enzyme of purine catabolism, to alloxanthine (oxipurinol) which remains tightly attached to the active site and so prevents the enzyme from functioning. In effect, allopurinol blocks uric acid production and causes the serum concentrations of hypoxanthine and xanthine to rise. Biochemically, gout is characterized by an overproduction of uric acid which, because of its low solubility, crystallizes in the lubricating fluid of the joints (synovial fluid) and becomes extremely painful. Allopurinol alleviates the condition by switching accumulation of uric acid to that of the much more soluble compound hypoxanthine which can be readily excreted. A second therapeutic effect of allopurinol is inhibition of purine biosynthesis and hence of purine availability. It is not entirely clear how allopurinol inhibits purine biosynthesis but there is evidence that it becomes ribotylated by the purine salvage mechanism and that this process may deplete stocks of PRPP (5-phosphoribosyl-1-pyrophosphate), the main biological ribotylation agent and an important purine precursor.

An important purine analogue in cancer chemotherapy is 6-mercaptopurine (23). This is ribotylated to the sulphydryl analogue of IMP and blocks amination of the latter to AMP. It also inhibits the *de novo* synthesis of purines. Azathioprine (Imuran) (24) was originally developed as a masked form of 6-mercaptopurine and used as a replacement for the latter in cancer chemotherapy but it was found to have a more selective effect than 6-mercaptopurine in blocking the immune response and has been widely used, since then, as an immunosuppressive for preventing rejection of transplanted organs, and in treating autoimmune diseases.

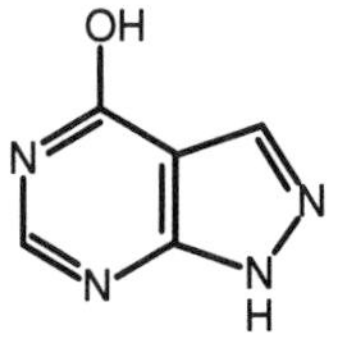

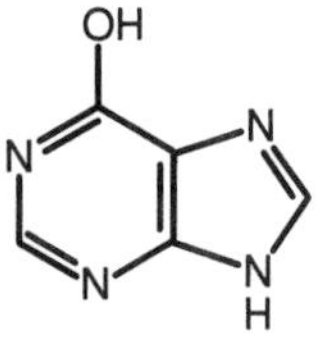

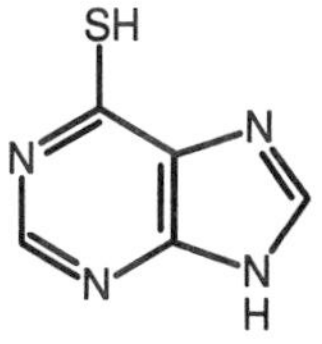

(23) 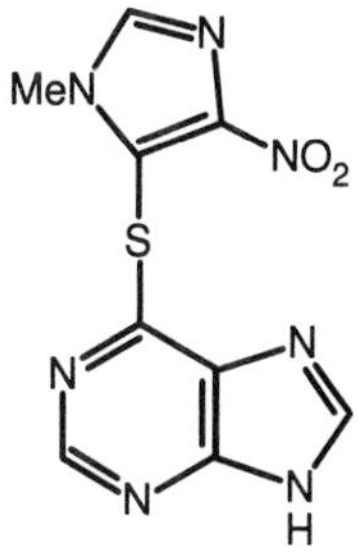 (24)

6.9 CONCLUSIONS

Purine derivatives, mainly in the form of ribotides, are ubiquitous and participate in every metabolic pathway. They are agents of energy transduction, redox catalysts, second messengers, and allosteric regulators. It is, then, all the more surprising that in contrast to their pyrimidine counterparts, purine derivatives and analogues feature only relatively infrequently in lists of natural or synthetic medicinal compounds and agricultural chemicals. Chemically, the *N*-glycosidic bond of purine ribosides is much more susceptible to hydrolysis than that of pyrimidine ribosides and biologically, both the catabolic and salvage pathways for purines are more in evidence than those for pyrimidines. This suggests that purine compounds, in general, do not make good pharmaceuticals or agricultural chemicals because of low persistence in a biochemical environment.

REFERENCES

Atkinson, D.E. (1977) *Cellular Energy Metabolism and its Regulation*, Academic Press, New York.

Brown, E.G. (1991) Purines, pyrimidines, nucleosides and nucleotides, in *Methods in Plant Biochemistry*, vol. 5 (ed L.J. Rogers), Academic Press, London, pp. 53–90.

Brown, E.G. and Konuk, M. (1995) Biosynthesis of nebularine (purine 9-β-D-ribofuranoside) involves enzymic release of hydroxylamine from adenosine. *Phytochemistry*, **38**, 61–71.

Buchanan, J.M., Sonne, J.C. and Delluva, A.M. (1948) Biological precursors of uric acid. *J. Biol. Chem.*, **173**, 69–79.

Eadie, J.S., Conrad, M., Toorchen, D. and Topal, M.D. (1984) Mechanism of mutagenesis by O^6-methylguanine. *Nature*, **308**, 201–203.

Greenberg, G.R. (1951) *J. Biol. Chem.*, **190**, 611–631.

Gulland, J.M. and Holiday, E.R. (1936) The constitution of the purine nucleosides. Part IV. Adenosine and related nucleotides and coenzymes. *J. Chem. Soc.*, 765–769.

Kenner, G.W., Taylor, C.W. and Todd, A.R. (1949) Experiments on the synthesis of purine nucleosides. Part XXIII. A new synthesis of adenosine. *J. Chem. Soc.*, 1620–1624.

Lijinsky, W. (1976) Interaction with nucleic acids of carcinogenic and mutagenic *N*-nitroso compounds. *Progr. Nucleic Acid Res. Mol. Biol.*, **17**, 247–269.

Lindahl, T. (1981) in *Chromosome Damage and Repair*, (eds E. Seeberg and K. Kleppe), Plenum Press, New York, p.207.

Nicholls, D.G. and Ferguson, S.J. (1992) *Bioenergetics II, An Introduction to the Chemiosmotic Theory*, Academic Press, London, pp. 190.

Roberts, J.J. (1975) in *Biology of Cancer*, (eds E.J. Ambrose and F.J.C. Roe), Ellis Horwood, Chichester.

Sutherland, E.W. (1971) in *Cyclic AMP*, by Robison, G.A., Butcher, R.W. and Sutherland, E.W., Academic Press, New York, pp. 13–15.

Zarbl, H., Sukumar, S., Arthur, A.V., Martin-Zanka, D. and Barbacid, M. (1985) Direct mutagenesis of Ha-*ras*-1 oncogenes by *N*-nitroso-*N*-methylurea during initiation of mammary carcinogenesis in rats. *Nature*, **315**, 382–385.

WIDER READING (BOOKS AND REVIEWS)

Adams, R.L.P., Knowler, J.T. and Leader, D.P. (eds) (1992) *The Biochemistry of the Nucleic Acids*, 11th edn, Chapman & Hall, London.

Fung, B-K., Hurley, J.B. and Stryer, L. (1981) Flow of information in the light-triggered cyclic nucleotide cascade of vision. *Proc. Nat. Acad. Sci.*, **78**, 152–156.

Lister, J.H. (1971) *Fused Pyrimidines, Part II, The Purines*, Wiley-Interscience, London.

Wald, G. (1968) The molecular basis of visual excitation. *Nature*, **219**, 800–807.

Wolstenholme, G.E.W. and O'Connor, C.M. (eds) (1957) *Chemistry and Biology of Purines*, Ciba Foundation Symposium, J & R Churchill, London.

CHAPTER 7

Pteridines

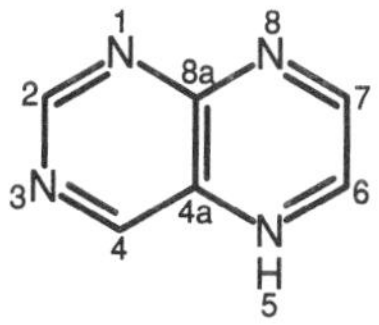

$$\text{[pteridine ring structure: positions 1 (N), 2, 3 (N), 4, 4a, 5 (NH), 6, 7, 8 (N), 8a]}$$

7.1 DISCOVERY AND NATURAL OCCURRENCE

Although Meldola drew attention in 1871 to the solubility in water of the yellow pigment of the common English brimstone butterfly (*Gonepteryx rhamni*) it was another twenty years before the pigment was examined chemically by Hopkins. His investigations into the pigments of butterfly wings revealed that there are several of these water-soluble compounds. Subsequently, they were shown to occur widely as insect eye pigments, and one, given the name xanthopterin, was discovered in human urine. It was not, however, until 1940 that the structures of these compounds began to be elucidated and confirmed by synthesis. This was largely the work of Wieland and Weygand and their respective associates. It was Wieland who, in 1941, named the compounds *'pteridines'* (Greek *pteron* = wing).

Since these early studies, pteridines have been isolated from a wide diversity of plant, animal and microbial sources. Probably the most familiar pteridine to the present day biochemist is the vitamin folic acid (Fig. 7.1). This is a bright yellow pigment, widely distributed in nature but amongst the richest dietary sources are green leaves, whence the name folic acid (Latin *folium* = leaf). Deficiency of this vitamin causes a macrocytic anaemia and has also been recently implicated as a major causative factor in *spina bifida*.

Folic acid was discovered during nutritional studies with microorganisms in the 1940s. A factor required from the growth of *Lactobacillus casei* and *Streptococcus faecalis* R was isolated and shown to be identical to a compound obtained from liver. In 1946, its structure was elucidated (Fig. 7.1) and described (Angier *et al., 1946)* in a scientific paper with sixteen co-authors!

Table 7.1 lists some of the more commonly occurring pteridines but for a more comprehensive account of their discovery and structure,

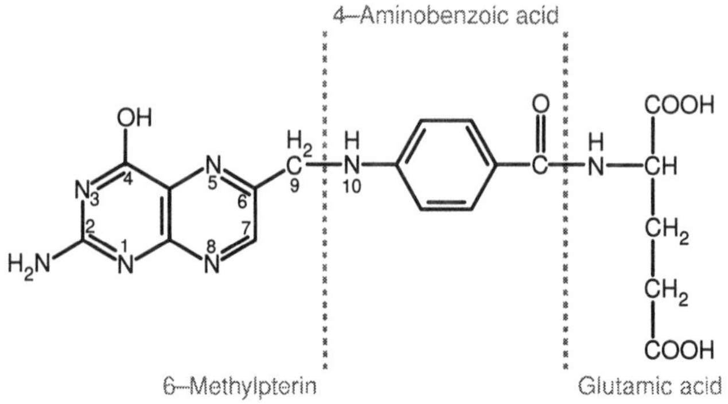

Fig. 7.1 Structure of folic acid showing the three components, *viz* 6-methylpterin, 4-aminobenzoic acid, and the amino acid glutamic acid. The term 'pterin' is given to 2-amino-4-hydroxypteridine, the most common substituted pteridine structure seen in nature.

the old but nevertheless excellent review by Gates (1941) is recommended.

7.2 CHEMICAL AND PHYSICAL PROPERTIES OF BIOCHEMICAL INTEREST

The pteridine ring system is that of a fused pyrimidine and pyrazine ring. The numbering of the rings has led to some confusion since earlier German and American systems conflicted with one another. Further complications have arisen from the adoption in biochemistry of the term 'pterin' for 2-amino-4-hydroxypteridine and is derivatives. Most naturally occurring pteridines have this substitution pattern, which corresponds to that of the purine guanine, and they are in consequence collectively called 'pterins' (see Table 7.1). In chemistry, it is necessary to

Table 7.1 Some naturally occurring pteridines, originally discovered in insect wings and eyes

Compound	Trivial name	Colour
2-Amino-4-hydroxypteridine	pterin	pale-yellow
2-Amino-4,6-dihydroxypteridine	xanthopterin	yellow
2-Amino-4,7-dihydroxypteridine	isoxanthopterin	yellow
2-Amino-4,6,7-trihydroxypteridine	leucopterin	colourless
7-Methylxanthopterin	chrysopterin	yellow
7-Propenylxanthopterin	erythropterin	orange-red
2-Amino-4-hydroxy-6[(L-erythro)-1′,2′-dihydroxypropylpterin]	biopterin	pale-yellow

number the two carbon atoms that are common to both rings but in biochemistry this has not, so far, been the case as there are no known natural products substituted at these sites. The simpler numbering system used in biochemistry is shown in Table 7.1 and Fig. 7.1. This is the system adopted in the present account.

Like the purines, the pteridines exhibit tautomerism. The low solubility of the compounds under physiological conditions, and their high melting points, suggest that as with their purine and pyrimidine counterparts, the keto forms predominate under biological conditions.

Virtually all pteridines are of low solubility in common organic solvents and in water. Polyhydroxy compounds such as ethylene glycol and glycerol do, however, appear to have some solvent power. Because of their low solubility, it is often difficult to crystallize pteridines and they tend to decompose rather than melt when heated.

Pteridines exhibit well-defined ultraviolet and visible absorption spectra with two or three major peaks and like the flavins, pteridines strongly fluoresce in ultraviolet light. Even simple derivatives exhibit this phenomenon, which led Kuhn to propose the name *lumazine* for 2,4-dihydroxypteridines. The colour of the fluorescence varies with pH, e.g. a solution of xanthopterin fluoresces red at a low pH but between pH 7–11 this becomes sky-blue.

Because the pteridine ring system contains four nitrogen atoms and each is situated in a six-membered ring, pteridine has a marked electron deficiency at each carbon site. This results in a lack of aromatic stability and favours ring-opening and addition reactions. There are, however, few pteridines in nature with strong electron-attracting substituents. Instead, the majority possess electron-releasing groups, such as hydroxy and amino, and they are, therefore, stable to ring opening. The parent compound *pteridine* is basic (pK_a = 4.12) and introduction of alkyl or amino groups increases this basicity whereas hydroxy substituents result in acidic compounds.

In contrast to the other families of compounds considered in this book, the chemical and physical properties of the pteridines have been much less thoroughly investigated and documented.

7.3 BIOSYNTHESIS OF FOLATES

As described in Chapter 8 (Figs. 8.3, 8.4) the biosynthesis of riboflavin involves the intermediate formation of a pteridine, 6,7-dimethyl-8-ribityl-lumazine. It is not surprising, therefore, to find that the main route of pteridine biosynthesis closely parallels that of riboflavin. Both biosyntheses are facets of the enzymic, THFA-linked, ring-opening of a purine and removal of C-8 (Fig. 7.2). It is interesting, in this respect, to note that tetrahydrofolic acid is involved in its own biosynthesis.

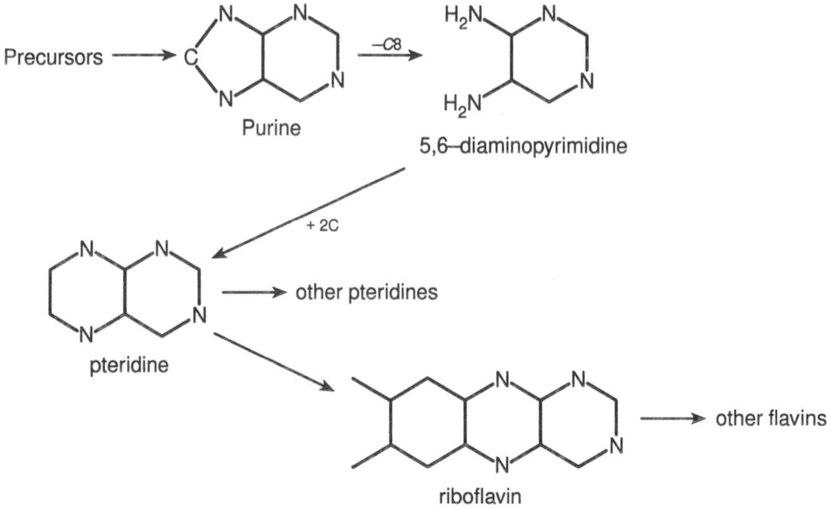

Fig. 7.2 Sequence of the main events in the linked biosynthesis of pteridines and flavins.

Fig. 7.3 shows the pathway of tetrahydrofolate formation. Like riboflavin biosynthesis, it begins with the enzymic opening of the imidazole ring of the purine nucleotide guanosine 5′-triphosphate and the transfer of C-8 to the receptor coenzyme tetrahydrofolic acid. This is followed by an Amadori rearrangement of the ribose moiety and cyclization to yield a partially reduced pteridine, dihydroneopterin triphosphate. These first three steps in the biosynthesis are all catalysed by the same enzyme, cyclohydrolase. Dephosphorylation and shortening of the side-chain of dihydroneopterin triphosphate then results in formation of hydroxymethyl dihydropterin with glyoxal as a byproduct of the chain-shortening. The ensuing step activates the hydroxymethyl group, by phosphorylation with ATP, so that the diphosphate condenses with 4-aminobenzoate yielding dihydropteroic acid and releasing inorganic pyrophosphate. Finally, dihydropteroic acid undergoes condensation with glutamic acid yielding dihydrofolic acid which is further reduced, enzymically, to tetrahydrofolic acid. The reductant is NADPH and the enzyme catalysing the reaction is dihydrofolate reductase.

There appears to be little information available about the origin of folic acid, the vitamin form of tetrahydrofolic acid, but it would seem probable that it arises from the oxidation of the latter or of dihydrofolic acid.

Other pteridines, such as the wing and eye pigments referred to in Section 7.1, originate from dihydroneopterin (Fig. 7.4, centre, top). In

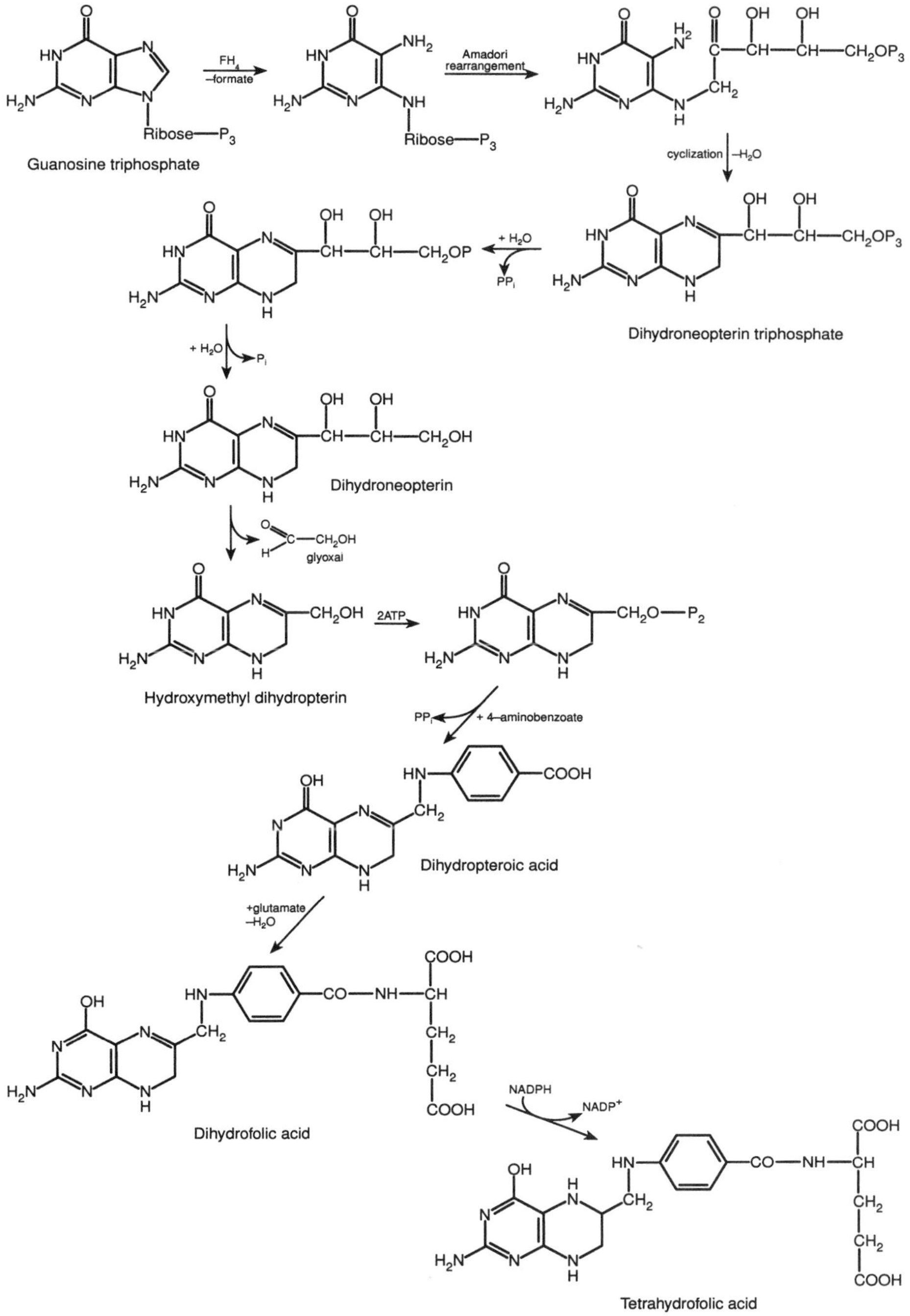

Guanosine triphosphate

FH4 –formate

Ribose—P3

Amadori rearrangement

cyclization –H2O

Dihydroneopterin triphosphate

+ H2O PPi

+ H2O Pi

Dihydroneopterin

glyoxal

Hydroxymethyl dihydropterin

2ATP

PPi + 4-aminobenzoate

Dihydropteroic acid

+glutamate –H2O

Dihydrofolic acid

NADPH NADP+

Tetrahydrofolic acid

Fig. 7.3 Pathway of tetrahydrofolic acid synthesis. Tetrahydrofolate is the coenzymic form of the vitamin folic acid. The enzyme GTP cyclohydrase catalyses the initial ring-opening reaction and requires tetrahydrofolic acid as its coenzyme.

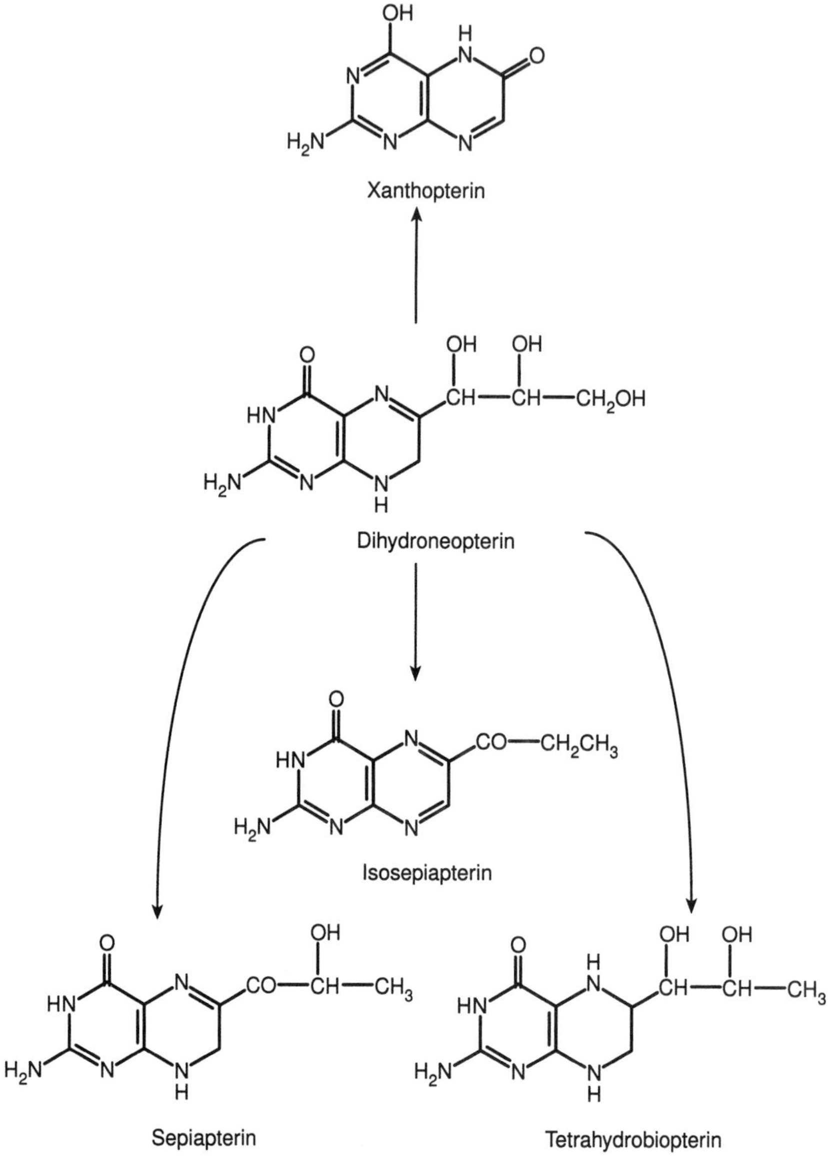

Fig. 7.4 Dihydroneopterin, an intermediate in folate biosynthesis (Fig. 7.3) is the precursor of the other biopterins found in nature.

the formation of tetrahydrobiopterin, sepiapterin, and isosepiapterin, the side-chain is retained (Fig. 7.4) whereas it is lost completely in the formation of xanthopterin, and hence of its oxidation product leucopterin (Table 7.1).

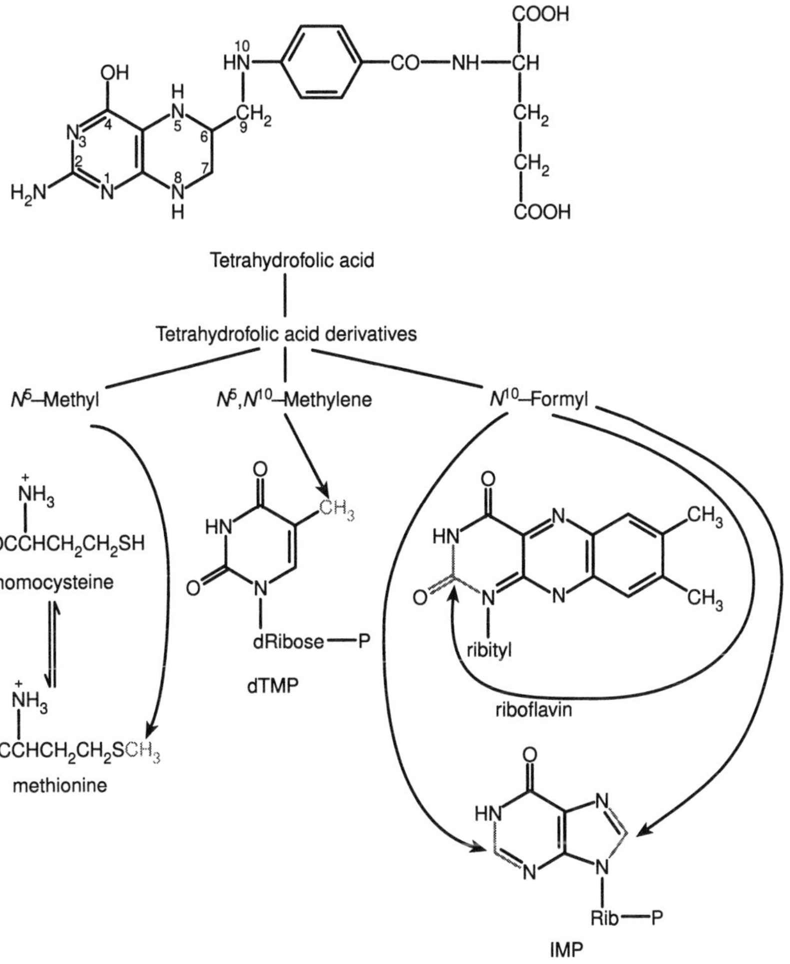

Fig. 7.5 Some examples of tetrahydrofolate-transported one-carbon units and their biosynthetic fate.

7.4 CATABOLISM

Little information is available concerning the catabolism of pteridine derivatives.

7.5 BIOCHEMICAL ROLE OF FOLIC ACID AND ITS DERIVATIVES

Folic acid (Fig. 7.1) is an essential, water-soluble, dietary factor and a member of the B-group of vitamins. Its metabolically functional form is tetrahydrofolic acid, the structure of which is shown in Fig. 7.3. There is evidence that the enzymes of folate metabolism show a preference for polyglutamate forms of folic acid in which the glutamate residue is condensed with other glutamate molecules to form a peptide.

At a cellular level, tetrahydrofolic acid is primarily concerned with the enzymic transfer of one-carbon units. Examples of this (Fig. 7.5) are seen in the metabolism of the amino acids methionine and histidine, the biosynthesis of purines, and of riboflavin and thymine. The one-carbon units commonly carried by tetrahydrofolic acid are methyl, methylene, and formyl, which correspond to the oxidation levels of methanol, methanal (formaldehyde) and formic acid. Whilst still attached to the carrier coenzyme, these one-carbon units can be enzymically modified (Fig. 7.6) e.g. N^5,N^{10}-methylene- can be reduced by NADPH to N^5-methyl tetrahydrofolate.

The main source of the one-carbon units transported by tetrahydrofolic acid is the serine hydroxymethyltransferase reaction (Fig. 7.7) during which serine is converted to glycine and N^5,N^{10}-methylenetetrahydrofolate is generated. In addition to its need for tetrahydrofolic

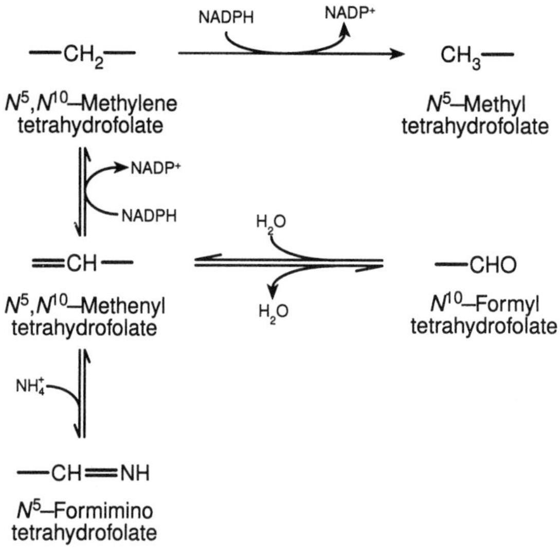

Fig. 7.6 Enzymic modification of one-carbon units attached to tetrahydrofolate.

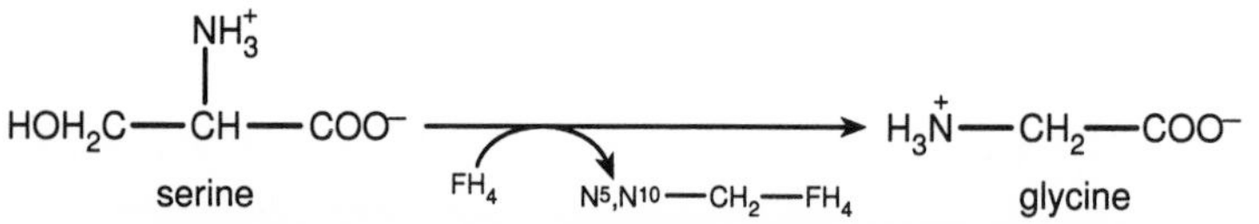

Fig. 7.7 The serine hydroxymethyltransferase reaction in which the coenzyme tetrahydrofolate (FH_4) accepts the transferred one-carbon unit to yield N^5,N^{10}-methylenetetrahydrofolate.

acid, the enzyme has a coenzyme requirement for pyridoxal 5′-phosphate.

Overt folic acid deficiency is seldom seen, probably because the intestinal flora synthesize sufficient of the vitamin to supplement the dietary intake. In expectant mothers living on a folate-minimal diet, the requirement of additional folate for foetal nucleic acid synthesis, and hence for protein formation, may well cause an overall deficiency. Recent research has implicated such folic acid deficiency in pregnancy, as a major factor in the birth defect *spina bifida*. Folic acid deficiency later in life is characterized by a macrocytic anaemia, again resulting from decreased protein synthesis. This anaemia is haematologically similar to pernicious anaemia but readily distinguishable from it biochemically.

As described above, there is evidence that the enzymes utilizing folic acid and tetrahydrofolate show a preference for polyglutamate forms of these compounds. In keeping with this, most dietary folate, coming as it does from plant and microbial sources, is in the polyglutamate form. There is, however, a problem. The folate polyglutamates cannot be absorbed from the intestine until all but one of the glutamate residues is hydrolytically removed. The enzyme catalysing this hydrolysis, γ-*L*-glutamyl carboxypeptidase is normally located in the brush border cells of the intestinal mucosa but in those diseases in which degeneration of the mucosa occurs, e.g. intestinal cancer, and tropical or non-tropical sprue, this enzymic activity is lost and hence there is a failure to absorb folate from the diet. Clinical studies have shown that whereas 1.5 mg of orally administered folate polyglutamates per day produces no haematological response, the monoglutamate alleviates the condition at doses as low as 25 μg per day.

In relation to the physiological functions of pteridines, two other compounds are worthy of mention. Biopterin (1) originally discovered as an insect pigment (Table 7.1) is widely distributed in the L-*erythro* form and functions as a growth factor for insects. It has also been found in human urine, extracts of the fruit-fly (*Drosophilia melanogaster*) and in queen bee jelly. Neopterin (2) is the second of these compounds. It exists in 4 isomeric forms (D-*erythro*-, L-*erythro*-, D-*threo*-, and L-*threo*-) but only two of these have been found in nature. The L-erythro form is widely distributed and like L-erythrobiopterin, functions as a growth factor for insects.

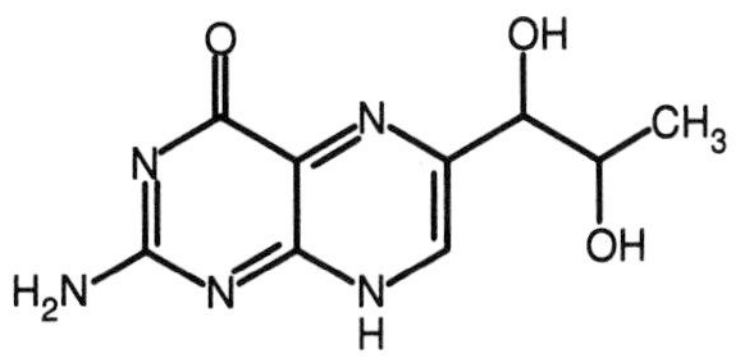

(1)

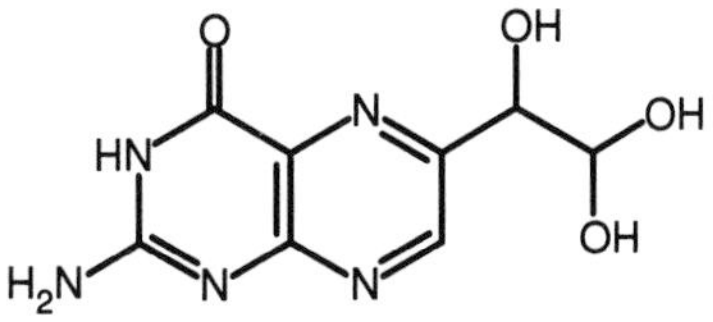

(2)

It is also a growth factor for the protozoan *Crithidia fasciculata* and has been found in the pupae of bees and in extracts of the bacterium *Serratia indica*.

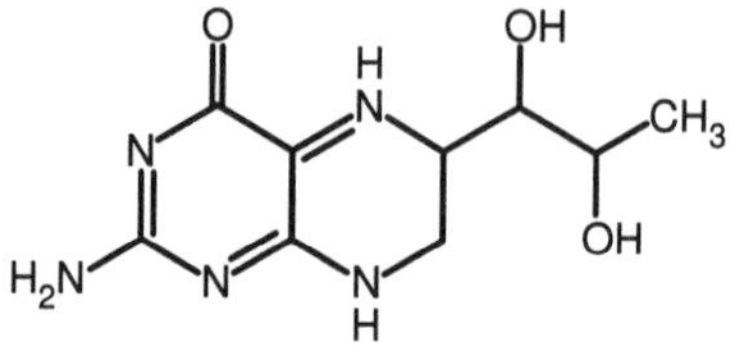

(3)

5,6,7,8-tetrahydrobiopterin (3), the biopterin analogue of tetrahydrofolic acid, is the coenzyme of phenylalanine hydroxylase. This enzyme is

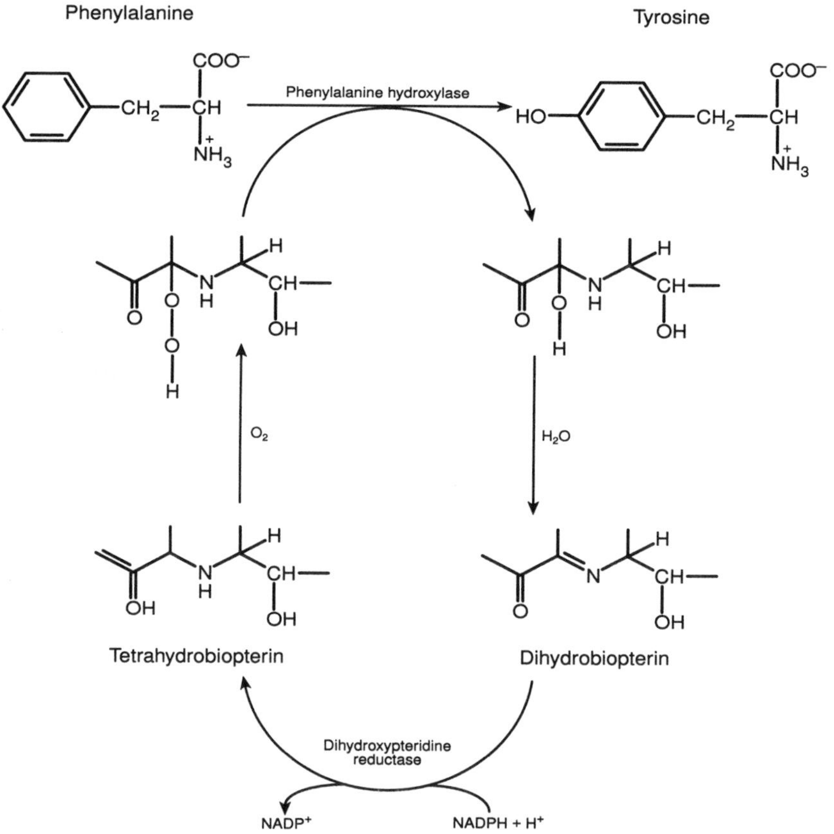

Phenylalanine

Tyrosine

Phenylalanine hydroxylase

O_2

H_2O

Tetrahydrobiopterin

Dihydrobiopterin

Dihydroxypteridine reductase

$NADP^+$

$NADPH + H^+$

Fig. 7.8 Hydroxylation of phenylalanine to tyrosine by the monooxygenase phenylalanine hydroxylase. The coenzyme tetrahydrobiopterin functions as the reductant and in behaviour typical of a monooxygenase, one atom of O_2 appears in the product and the other in H_2O. Dihydrobiopterin is enzymically reduced by a reductase using NADPH, so regenerating tetrahydrobiopterin.

a mixed-function oxygenase, i.e. a monooxygenase, catalysing the hydroxylation of phenylalanine to tyrosine (Fig. 7.8), an essential step in phenylalanine catabolism. Absence or deficiency of phenylalanine hydroxylase, or more rarely of the pterin coenzyme, causes an accumulation of phenylalanine in the tissues and leads to the disease phenylketonuria, so-called because of the excretion of the phenylketo acid phenylpyruvic acid. The latter originates from the metabolism of the accumulating phenylalanine by an aminotransferase.

As can be seen in Fig. 7.8, the reaction catalysed by phenylalanine hydroxylase requires O_2, and tetrahydrobiopterin serves as the reductant. Typical of a monooxygenase, and as shown in Fig. 7.8, one of the atoms of oxygen from the O_2 ends up in the product, and the other in the molecule of water released. During the hydroxylation process, coenzymic tetrahydrobiopterin is oxidized to dihydrobiopterin and has to be regenerated by dihydropterin reductase using NADPH as the reductant (Fig. 7.8).

7.6 SYNTHETIC COMPOUNDS OF CLINICAL IMPORTANCE

Because of the essential involvement of tetrahydrofolate as a catalyst in the *de novo* biosynthesis of nucleotides, and hence of nucleic acids and proteins, numerous attempts have been made, with varying degrees of success, to design folate antagonists for use in cancer chemotherapy. The earliest of these, first synthesized and tested in the late 1940s, were aminopterin (4) and its N^{10}-methyl derivative, known as methotrexate (5). These are analogues of folic acid and N^{10}-methylfolate respectively.

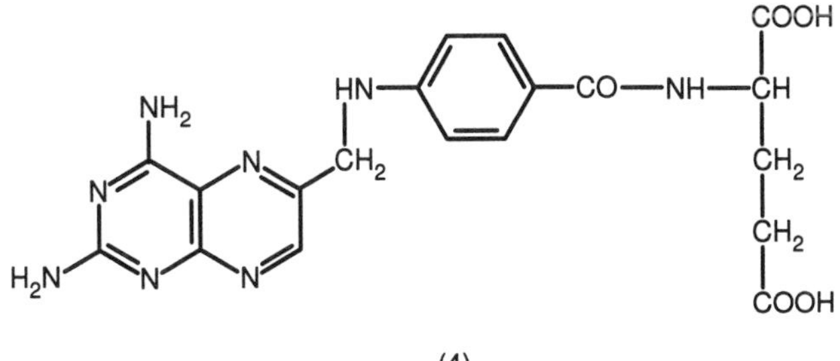

(4)

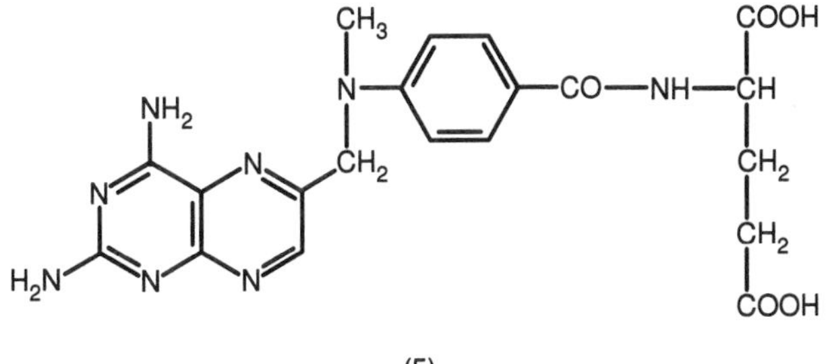

(5)

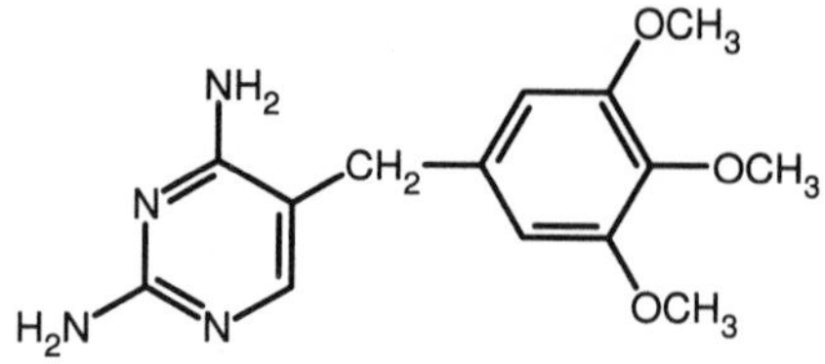

(a)

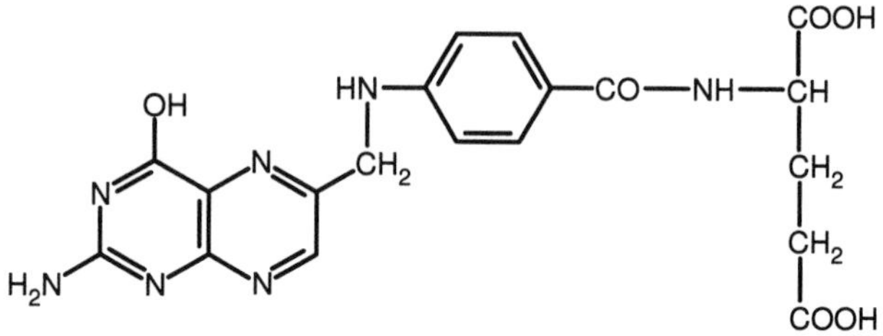

(b)

Fig. 7.9 Structural relationship of (a) trimethoprim to (b) folic acid. Trimethoprim is a potent inhibitor of dihydrofolate reductase and is used clinically as an antibacterial and antiprotozoal drug.

Although the alteration to chemical structure represented by these analogues is minimal, an amino replacing the 4-hydroxyl group in both cases, the biochemical effects are profound. The two compounds are powerful inhibitors of dihydrofolate reductase, the enzyme catalysing the reduction of folate to dihydrofolate, and of dihydrofolate to tetrahydrofolate. In consequence, they prevent conversion of dietary folate to coenzymic tetrahydrofolate, and regeneration of tetrahydrofolate from dihydrofolate.

Methotrexate is a valuable anticancer drug and is widely used in the treatment of rapidly growing cancers, such as acute leukaemia and choriocarcinoma. Unfortunately, because it kills all rapidly dividing cells irrespective of whether they are normal or abnormal, the drug adversely affects the epithelial cells of the intestinal tract, the hair follicles, and the stem cells of the bone marrow. This gives rise to the characteristic toxicity syndrome associated with cancer chemotherapy.

Trimethoprim is a synthetic antibacterial pyrimidine derivative used widely in clinical medicine. This, too, functions as an inhibitor of dihydrofolate reductase and could be regarded as a folate analogue (Fig. 7.9).

REFERENCES

Angier, R.B. *et al.* (1946) *Science*, **103**, 667–671.

WIDER READING (BOOKS AND REVIEWS)

Bacher, A., Mailänder, B., Baur, R., Eggers, V., Harders, H. and Schnepple, H. (eds) (1975) *Chemistry and Biology of Pteridines*, de Gruyter, Berlin.

Brown, G.M. and Williamson, J.M. (1982) Biosynthesis of riboflavin, folic acid, thiamine and pantothenic acid, *Advances in Enzymology*, **53**, 345–381.

Gates, M. (1947) The chemistry of the pteridines, *Chemical Reviews*, **41**, 53–95.

Shiota, J. (1971) Biosynthesis of folic acid and 6-substituted pteridines, in *Comprehensive Biochemistry*, (eds M. Florkin and E.H. Stotz), Vol. 21, Elsevier, Amsterdam, pp. 111–152.

Shive, W. (1963) Folic acid and pteridines, in *Comprehensive Biochemistry* (eds M. Florkin and E.H. Stotz), Vol. 2, Elsevier, Amsterdam, pp. 83–102.

CHAPTER 8

Flavins

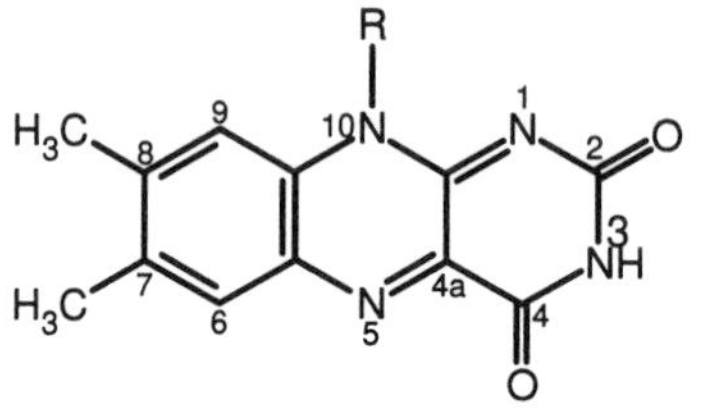

8.1 DISCOVERY AND NATURAL OCCURRENCE

The flavins are a widely distributed group of yellow pigments of prime biochemical importance. Chemically, they are derivatives of isoalloxazine (1) and as such, are modified pteridines; more specifically, they are benzopteridines. The parent compound of the group, riboflavin (2), was initially discovered as an essential dietary factor for rat growth. It was first isolated, in 1933, by Györgyi, Wagner-Jauregg and Kuhn. With the cooperation of a Bavarian cheese dairy, they started with 5400 litres of whey, from which they eventually obtained 1 g of a yellow-orange crystalline compound that they designated *lactoflavin* (L. *lactis* = milk, *flavis* = yellow). Almost simultaneously, another group obtained a similar compound from eggs and called it *ovoflavin*. Later, a third compound with similar properties was extracted from liver and called *hepatoflavin*. It was, however, soon realized that these three compounds are one and the same thing. In recognition of the detectable presence in the com-

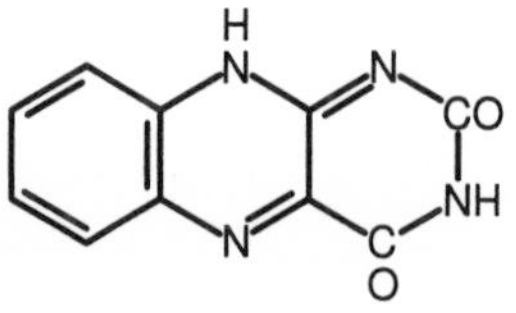

(1)

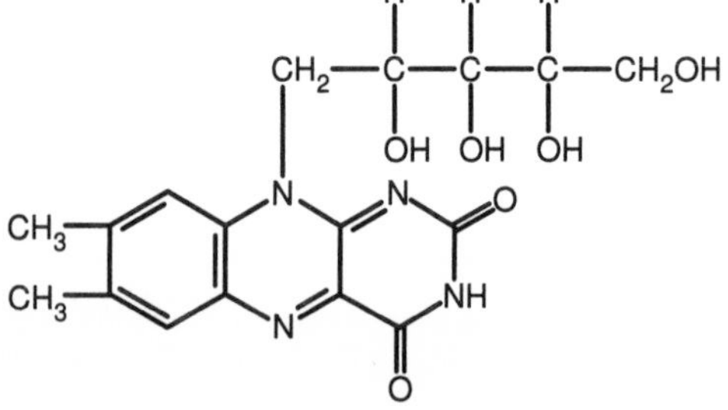

(2)

pound of a ribose-related sugar residue, later identified as ribitol, the compound subsequently became known as riboflavin.

Relatively high concentrations of riboflavin are found in the eyes of many species of fish, and of crabs, and also in the Malpighian tube of fireflies and cockroaches (the concentration in the latter being 40 times higher than that in beef liver). Some fungi produce large amounts of riboflavin. Of particular interest in this respect is *Eremothecium ashbyii*, discovered as a cotton parasite that stains the cotton bolls of infected plants bright-yellow. This colour is, in fact, riboflavin and such a large amount is produced by the mould that it crystallizes inside the mycelium with large yellow crystals being clearly visible under the microscope. Industrial fermentations with *E. ashbyii* are used to produce riboflavin for pharmaceutical purposes. Another report of high local concentrations of riboflavin concerns the eye of the lemur; more specifically, the *tapetum lucidum* (the reflective inner lining of the eye) of this animal has been shown to consist of a layer of crystalline riboflavin.

Most riboflavin in biological systems is not present in a free state but either as its 5′-phosphate, *flavin mononucleotide* (3), known as FMN, or in combination with AMP as *flavin adenine dinucleotide* (FAD) the structure of which is shown in (4). For example, the major part of the riboflavin present in human milk is present as FAD; and almost all of the riboflavin present in rat kidneys is in the form of FMN and FAD.

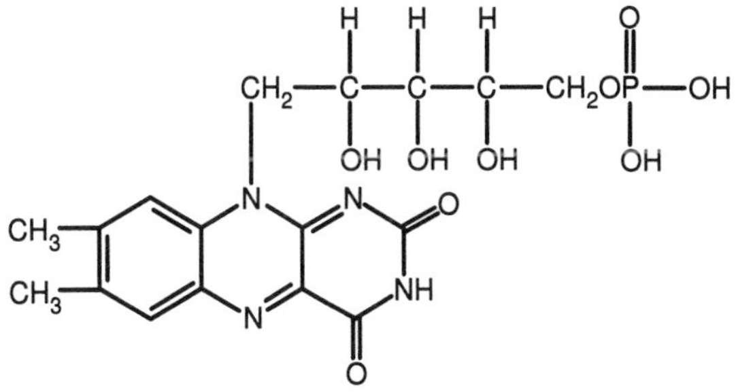

(3)

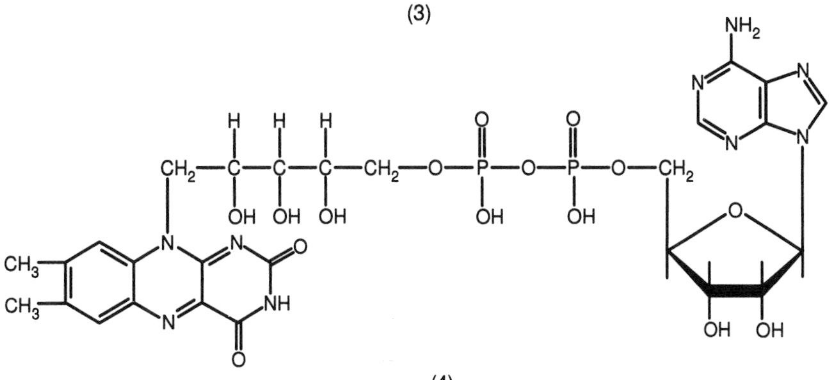

(4)

8.2 CHEMICAL PROPERTIES OF BIOCHEMICAL INTEREST

Riboflavin is essentially water-soluble but only to the extent of 10–13 mg in 100 ml at room temperature. It is less soluble in ethanol, slightly soluble in higher alcohols, and insoluble in most common organic solvents. Clinically, the sodium salt of riboflavin 5′-phosphate which is highly soluble in aqueous media is used for riboflavin injections.

In neutral, aqueous solutions, riboflavin exhibits an intense yellow-green fluorescence which disappears on addition of acid or alkali. The ultraviolet-visible absorption spectrum of riboflavin in water at biological pH values shows characteristic absorption maxima at 223, 268, 359–375, 446, and 475 nm, and spectrophotometry therefore affords a good means of quantitative analysis.

Sodium dithionite ($Na_2S_2O_4$) and similar reducing agents reduce riboflavin, 2 hydrogens per molecule being taken up to yield a colourless *leuco* compound. Catalytic hydrogenation produces octahydroflavins which, in alkaline solution, are readily reoxidized by air to hexahydroflavins.

Solutions of riboflavin are unstable in bright light, the nature of the products being dependent on the pH of the solution. Irradiation under alkaline conditions brings about loss of the distal 4 carbon atoms of the ribitol side-chain, leaving the residual carbon atom as a methyl group. The product of this photodegradation (Fig. 8.1) is lumiflavin which, unlike riboflavin, can be extracted into chloroform. In neutral or acidic solutions, the entire ribitol side-chain is lost by photodegradation, yielding a chloroform-soluble alloxazine known as lumichrome (Fig. 8.1). Lumiflavin, like riboflavin, exhibits a yellow fluorescence in ultraviolet light whereas under similar conditions lumichrome has a bright sky-blue fluorescence. The properties are of use in detecting the compounds on chromatography plates.

Formation of molecular complexes between flavins and purines such as adenine and caffeine have been reported to occur. Similarly, indoles, e.g. tryptophan and serotonin, can complex with flavins. Chlortetracycline and various phenols also exhibit this property in respect of flavins.

8.3 BIOGENESIS OF FLAVINS AND PTERIDINES

As pointed out in Chapter 6 (Section 6.4), the pathways of purine, pteridine, and flavin biosynthesis are closely connected. Structurally, pteridines can be regarded as ring-expanded purines; and riboflavin biogenesis involves the intermediate formation of a pteridine. Fig. 6.5 in Chapter 6 summarizes these biosynthetic relationships.

Early studies with the strongly flavinogenic fungus *Eremothecium*

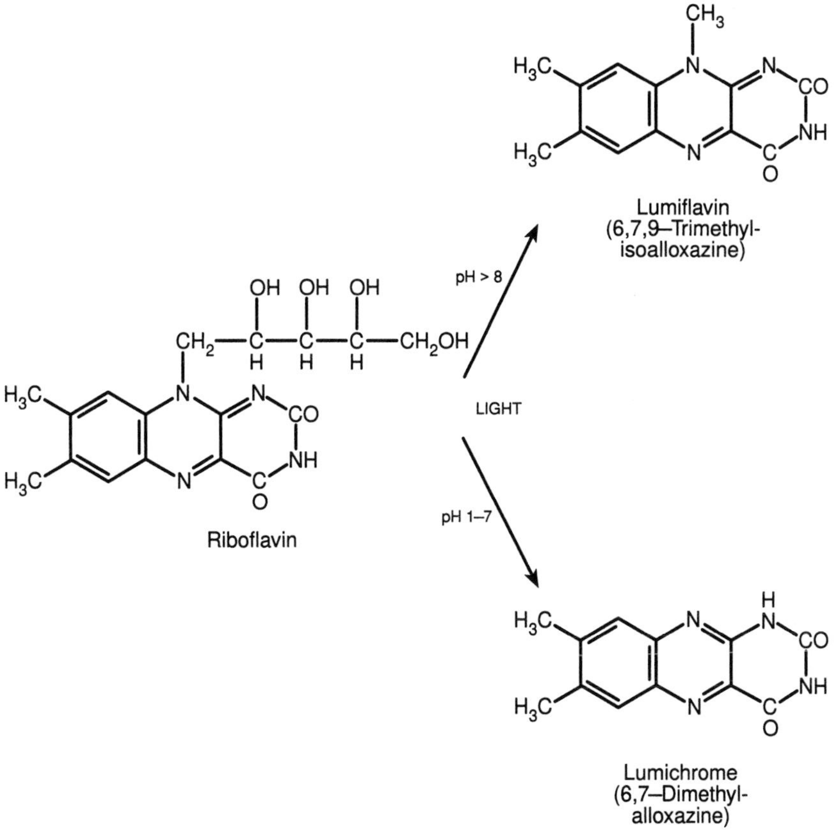

Fig. 8.1 The photochemical degradation of riboflavin. Under acidic or neutral conditions the product is lumiflavin whereas in alkaline solutions, lumichrome is formed.

ashbyii showed that purines specifically stimulate riboflavin synthesis (MacLaren, 1952; Goodwin and Pendlington, 1954; Brown *et al.*, 1955; Brown *et al.*, 1958). Later, a remarkable similarity was seen in the labelling pattern in analogous parts of the purine, flavin and pteridine ring systems when *E. ashbyii*, or, in the case of the pteridines, butterflies' wings were supplied, separately, with ^{14}C-labelled formate, $^{14}CO_2$ and ^{14}C-labelled glycine. Studies using ^{15}N-glycine and glycine dual-labelled with ^{14}C and ^{15}N revealed that glycine is incorporated as an intact unit into comparable positions in all three ring systems. These observations are summarized in Fig. 8.2.

Subsequent work with [U-^{14}C]adenine demonstrated a substantial uptake of radioactivity into riboflavin. When, however, [8-^{14}C]adenine was supplied instead, no significant amount of radioactivity was incorporated. The data obtained indicated that incorporation of adenine into

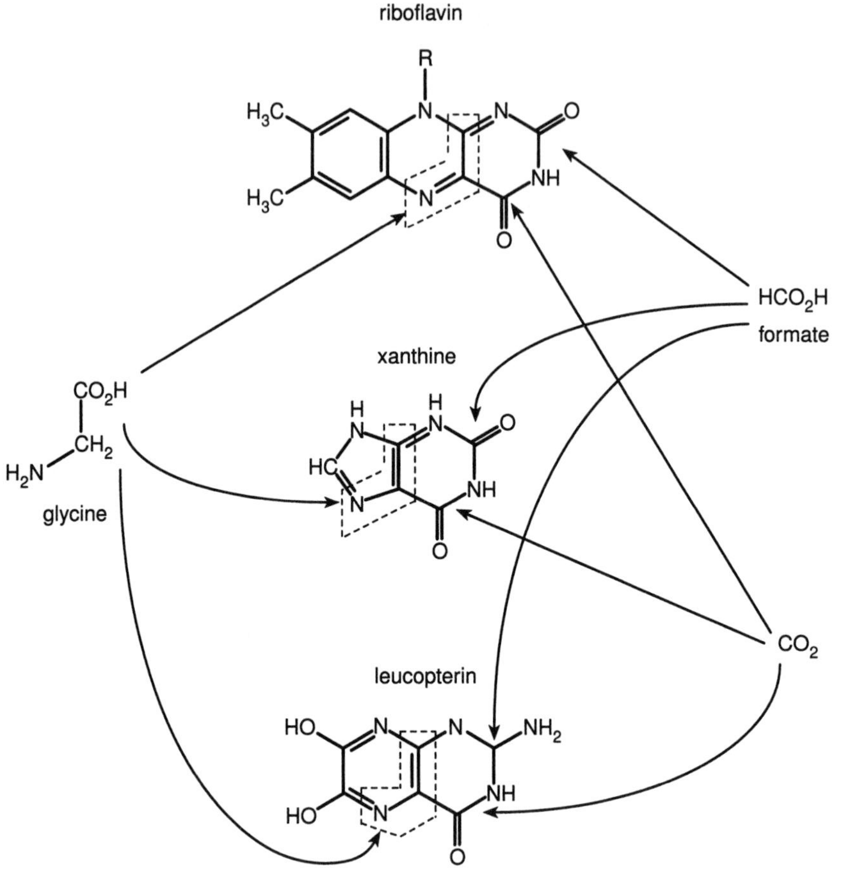

Fig. 8.2 Incorporation of ^{14}C-labelled glycine, formate and CO_2 into riboflavin, xanthine and leucopterin.

riboflavin involves initial loss of C-8 from the purine. As this is the ureido C-atom of the imidazole ring, its removal implies that a 5,6-diaminopyrimidine is intermediate between the purine precursor and riboflavin. This was subsequently confirmed.

8.3 BIOSYNTHESIS OF RIBOFLAVIN

Current knowledge of the biosynthetic pathway for riboflavin is shown schematically in Figs. 8.3 and 8.4. The starting point is guanosine 5′-triphosphate (GTP) which is in effect the link compound between purine biosynthesis and flavinogenesis. In a reaction catalysed by the enzyme GTP-cyclohydrolase and common to both riboflavin and folic acid biosynthesis, the imidazole ring of the purine nucleotide is opened and C-

8, the ureido carbon, is transferred, at the oxidation level of formate, to the coenzyme tetrahydrofolic acid. The product of the ring-opening reaction is the corresponding 5,6-diaminopyrimidine-5′-ribotide (Fig. 8.3). In some organisms, such as the mould *Ashbya gossypii*, the ribose moiety is then enzymically reduced to ribitol and the pyrimidine ring deaminated, hydrolytically, at the 2-position (Fig. 8.3). In other organisms, e.g. *E. coli*, the sequence of these two reactions is reversed, deamination preceding reduction. In both cases, the end product is 5-amino-2,4-dihydroxy-6-ribitylaminopyrimidine phosphate (Fig. 8.3). This is ring-closed by a 4C-unit derived from the ribitol phosphate residue of a second molecule of the same compound. At the same time, the phosphate group is lost, giving 6,7-dimethyl-8-ribityl-lumazine, the immediate precursor of riboflavin (Masuda, 1956).

Conversion of the lumazine derivative to riboflavin is a similar type of mechanism (Fig. 8.4), with two molecules of the substrate being needed to produce one molecule of riboflavin (Plaut, 1960; Goodwin and Horton, 1961). This enzymic reaction is catalysed by riboflavin synthase and results in the concomitant release of a dephosphorylated molecule of the earlier intermediate 5-amino-2,4-dihydroxy-6-ribitylaminopyrimidine. Radioisotopic evidence for this enzymic mechanism is shown in Fig. 8.4.

Interestingly, the riboflavin synthase reaction is extremely rapid and even crude extracts of *E. ashbyii*, obtained by grinding mycelium in a cold buffer solution, convert colourless 6,7-dimethyl-8-ribityl-lumazine into bright yellow riboflavin in a matter of seconds.

8.4 BIOCHEMICAL ROLE OF RIBOFLAVIN AND ITS DERIVATIVES

Riboflavin is a water-soluble vitamin of the B group and precursor of the two important flavin redox coenzymes flavin mononucleotide and flavin adenine dinucleotide. Flavin mononucleotide (FMN) is simply the 5′-monophosphate of riboflavin (2), whereas the molecule of flavin adenine dinucleotide (FAD) consists of a molecule of FMN linked, through its 5′-phosphate group to the 5′-phosphate group of adenosine monophosphate. In other words, two nucleotides linked tail-to-tail to give a dinucleotide (4). The formation of FMN and FAD from riboflavin is shown in Fig. 8.5.

FMN functions as an electron acceptor during the operation of mitochondrial NADH dehydrogenase. This enzyme reoxidizes NADH, produced during cellular respiration, and simultaneously reduces FMN:

$$NADH + H^+ + FMN \rightarrow FMNH_2 + NAD^+$$

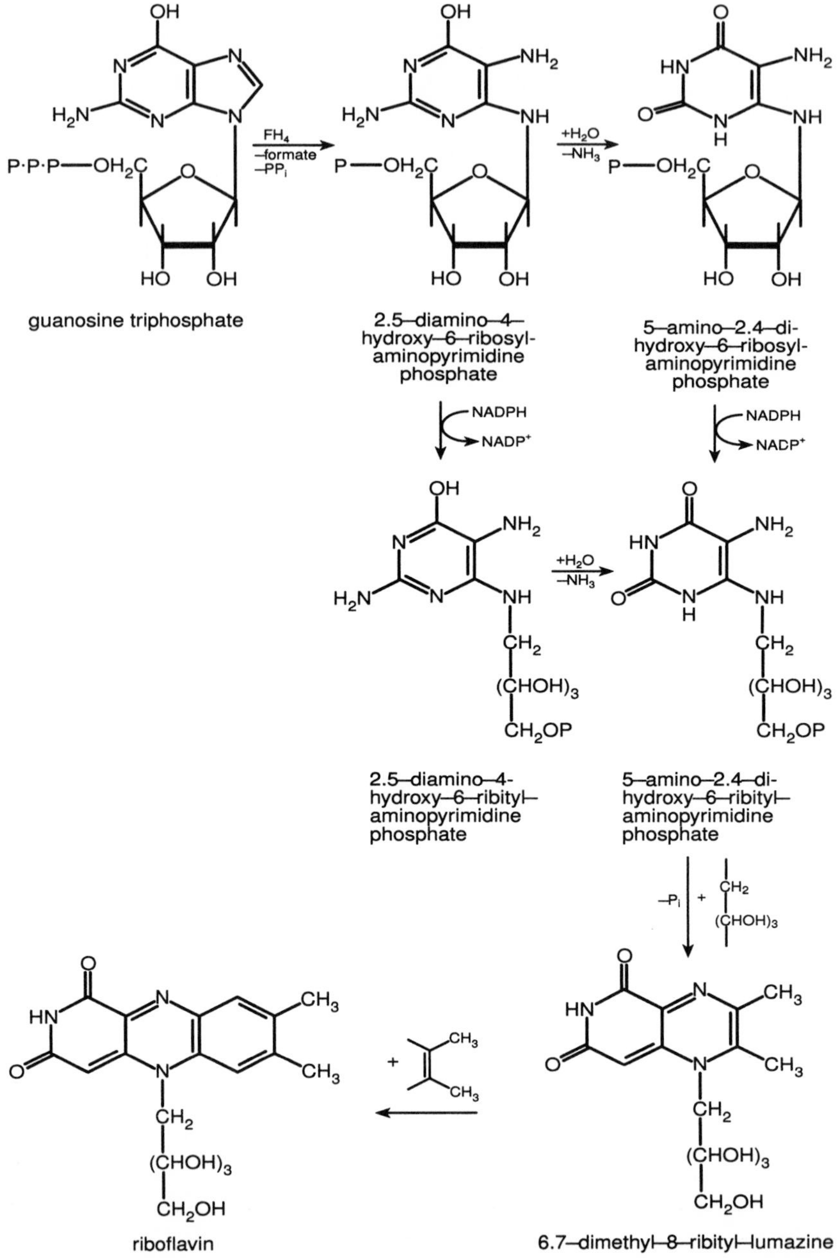

guanosine triphosphate

2.5–diamino–4–hydroxy–6–ribosyl-aminopyrimidine phosphate

5–amino–2.4–di-hydroxy–6–ribosyl-aminopyrimidine phosphate

2.5–diamino–4-hydroxy–6–ribityl–aminopyrimidine phosphate

5–amino–2.4–di-hydroxy–6–ribityl–aminopyrimidine phosphate

riboflavin

6.7–dimethyl–8–ribityl–lumazine

Fig. 8.3 Biosynthesis of riboflavin. The first reaction requires tetrahydrofolic acid as a coenzyme and C-8 is removed as a one-carbon unit. Some organisms (e.g. *Ashbya gossypii*) reduce the pyrimidine ribotide first and then deaminate. Others (e.g. *E. coli*) deaminate first, and then reduce the sugar. In the penultimate step, the 4-C unit derives from the sugar unit of a second molecule. The final step is shown in more detail in Fig. 8.4.

2 molecules

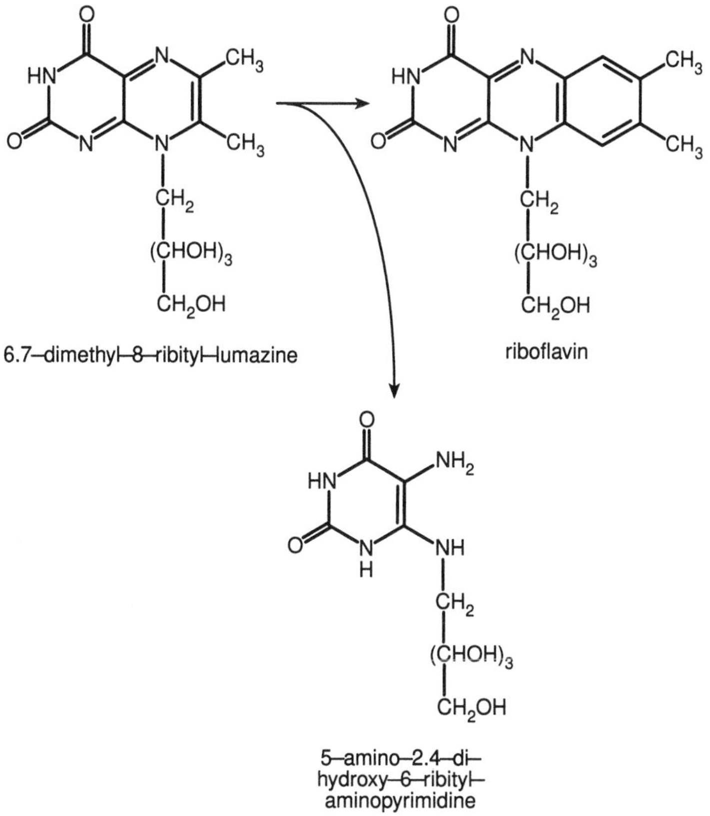

6.7–dimethyl–8–ribityl–lumazine

riboflavin

5–amino–2.4–di–hydroxy–6–ribityl–aminopyrimidine

Fig. 8.4 The enzymic conversion of 6,7-dimethyl-8-ribityllumazine to riboflavin, catalysed by riboflavin synthase, involves two molecules of the substrate per molecule of product. Radiolabelling experiments show that the substrate serves as donor and receptor of a 4-C unit, yielding a molecule of riboflavin and a molecule of the intermediate 5-amino-2,4-dihydroxy-6-ribitylaminopyrimidine (see Fig. 8.3). The ^{14}C-labelled 4-C unit and its fate is indicated by the shaded parts of the structures.

The dinucleotide FAD also plays an important part in a number of biochemical oxidation-reduction processes. It is the prosthetic group (i.e. the tightly bound coenzyme) of succinate dehydrogenase, a component enzyme of the tricarboxylic acid cycle, and has a similar role in fatty acyl-CoA dehydrogenase in fatty acid metabolism. Another example is seen with the mitochondrial enzyme glycerol phosphate dehydrogenase, of which FAD is also the prosthetic group.

Although the complete reduction of a molecule of FMN, or that of FAD, involves 2-electrons, the process occurs in two single-electron

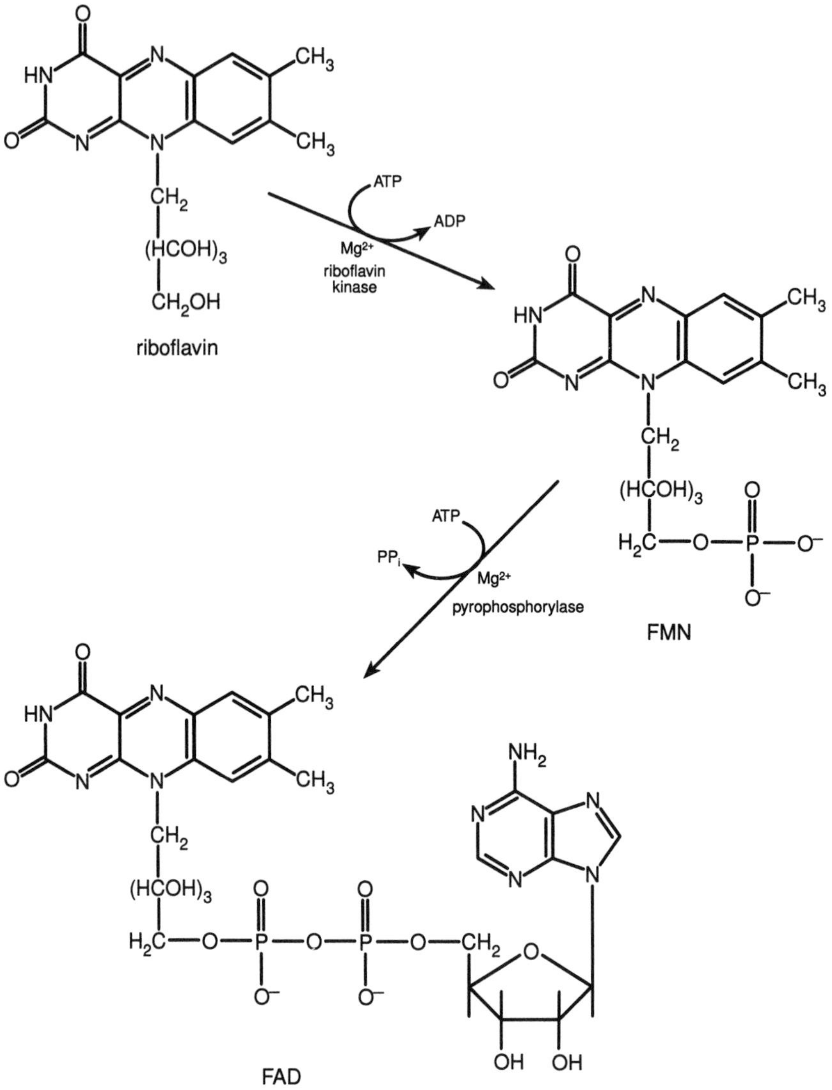

riboflavin

Mg^{2+} riboflavin kinase

ATP ADP

FMN

ATP PP_i Mg^{2+} pyrophosphorylase

FAD

Fig. 8.5 Formation of flavin mononucleotide (FMN) and flavin adenine dinucleotide (FAD) from their vitamin precursor riboflavin.

steps with formation of a semiquinone intermediate (Fig. 8.6). This means that reduced FAD or FMN can be reoxidized by one-electron acceptors. These various oxidation states of the flavins can be detected spectrophotometrically and whereas the oxidized compounds in aqueous solution are bright yellow (λ_{max} 450 nm) and the fully reduced ones are colourless, the semiquinone is red (λ_{max} 490 nm) or, when protonated, blue (λ_{max} 560 nm).

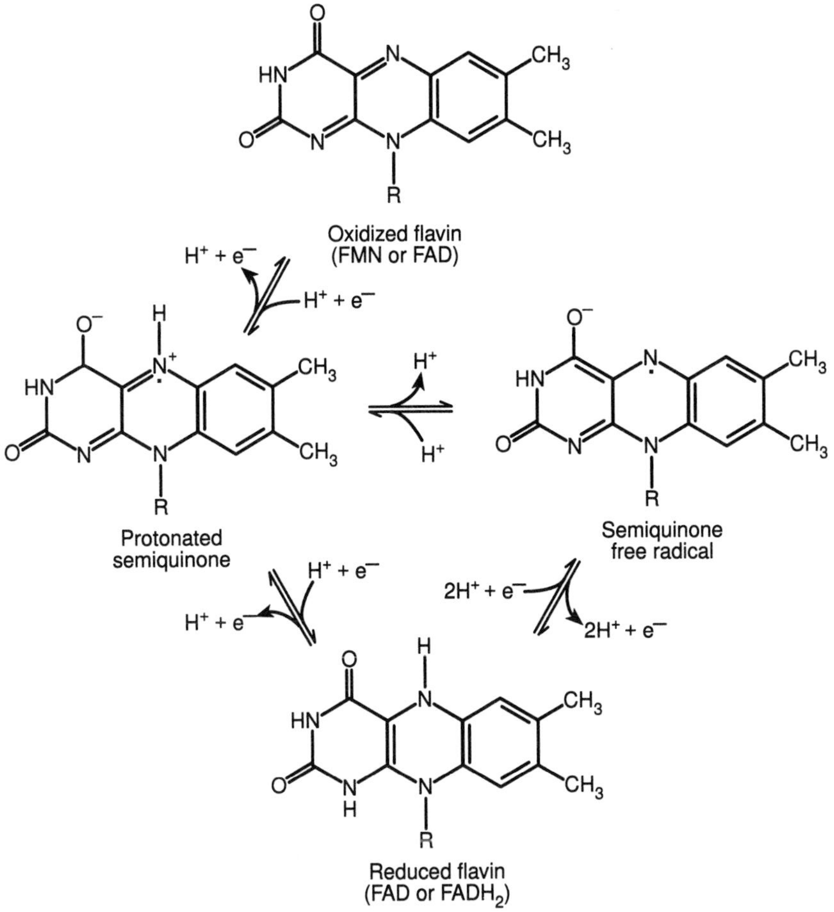

Oxidized flavin
(FMN or FAD)

Protonated
semiquinone

Semiquinone
free radical

Reduced flavin
(FAD or $FADH_2$)

Fig. 8.6 Oxidation states of FAD and FMN. Flavins participate in two-electron reactions but the formation of an intermediate semiquinone free radical permits these reactions to proceed one electron at a time. Reduced flavins can, in consequence, be reoxidized by one-electron acceptors.

8.5 SYNTHETIC COMPOUNDS OF MEDICINAL OR AGRICULTURAL INTEREST

There are few flavin analogues that fall into this category and only two worthy of mention here. Both of them are analogues of riboflavin in which the sugar moiety is modified. In one, the ribitol side-chain of riboflavin is replaced by the corresponding galactose derivative, galactitol. This modified flavin (5), known as galactoflavin, is a powerful antagonist of riboflavin and induces congenital deformations in animals. The related compound lyxoflavin (6) is the L-lyxose analogue of riboflavin. It is also a riboflavin antagonist yet has been used as a growth-pro-

(5) (6)

moting agent in animal feedstuffs. The apparent paradox is explained by the antimicrobial activity of the compound which would suppress growth of the gut flora and hence its uptake of nutrients, making more available to the host animal. A similar, but not to be encouraged, use has been made in some countries of antibiotic supplements in animal feedstuffs.

REFERENCES

Brown, E.G., Goodwin, T.W. and Jones, O.T.G. (1958) Studies on the biosynthesis of riboflavin. Purine metabolism and riboflavin synthesis in *Eremothecium ashbyii. Biochemical Journal*, **68**, 40–49.

Brown, E.G., Goodwin, T.W. and Pendlington, S. (1955) Studies on the biosynthesis of riboflavin. Further observations on nitrogen metabolism and flavinogenesis in *Eremothecium ashbyii. Biochemical Journal*, **61**, 37–46.

Goodwin, T.W. and Horton, A.A. (1961) Biosynthesis of riboflavin in cell-free systems. *Nature*, **191**, 772–774.

Goodwin, T.W. and Pendlington, S. (1954) Studies on the biosynthesis of riboflavin. Nitrogen metabolism and flavinogenesis in *Eremothecium ashbyii. Biochemical Journal*, **57**, 631–641.

MacLaren, J.A. (1952) The effects of certain purines and pyrimidines upon the production of riboflavin by *Eremothecium ashbyii. Journal of Bacteriology*, **63**, 233–241.

Masuda, T. (1956). G Compound isolated from the mycelium of *Eremothecium ashbyii. Pharm. Bulletin (Japan)*, **4**, 375–381.

Plaut, G.W.E. (1960) Studies on the stoichiometry of the enzymic conversion of 6,7-dimethyl-8-ribityllumazine to riboflavin. *Journal of Biological Chemistry*, **235**, PC 41–42.

WIDER READING (BOOKS AND REVIEWS)

Brown, G.M. and Williamson, J.M. (1982) Biosynthesis of riboflavin, folic acid, thiamine, and pantothenic acid. *Advances in Enzymology*, **53**, 345–381.

Demain, A.L. (1972) Riboflavin oversynthesis. *Annual Review of Microbiology*, **26**, 369–388.

Plaut, G.W.E. (1971) The biosynthesis of riboflavin, in *Comprehensive Biochemistry* (eds M. Florkin and E.H. Stotz), Vol. **21**, Elsevier, Amsterdam, pp. 11–44.

Sebrell, W.H. Jr. and Harris, R.W. (eds) (1972) *The Vitamins*, 2nd edn, Vol. 5, Chapter 14 *Riboflavin*, Academic Press, London, pp. 2–87.

CHAPTER 9

Indoles

9.1 DISCOVERY AND NATURAL OCCURRENCE

The indole unit is widely distributed in nature. Well over a thousand indole alkaloids have been described, many of major pharmacological importance. A majority of these alkaloids are of complex structure but have their biosynthetic origin in the simple indole amino acid tryptophan (1) which, itself, is an important cellular constituent. The vat dye indigo (2) is also an indole and has been known and widely used since antiquity. It was prepared by crushing the leaves of *Isatis tinctora* (woad) or *Indigofera tinctora* and allowing them to ferment. The resulting hydrolysis of plant indican[1] (indol-3-yl β-glucoside; 3) liberates the aglycone which is then enzymically oxidised by atmospheric oxygen to yield indigo (2). Today, indigo is still an important dyestuff but is of synthetic origin. Indole (4) the parent compound of the series was first isolated, from indigo, and its structure elucidated by von Bäyer. In highly dilute solutions, indole has a pleasant odour and is used in perfumery but in marked contrast, skatole (3-methylindole; 5) is responsible for the characteristically unpleasant smell of faeces.

(1)

(2)

[1] Unfortunately, some confusion arises over this name. Indican isolated from mammalian sources, such as urine, is indol-3-yl sulphate.

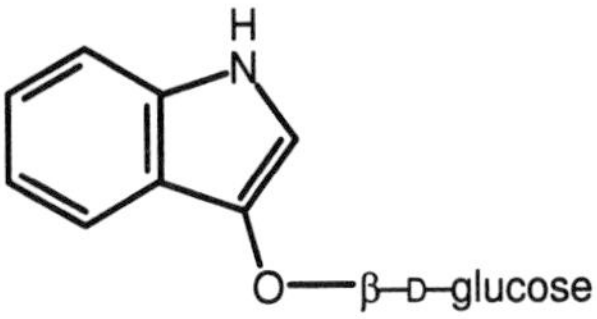

O—β-D-glucose

(3) 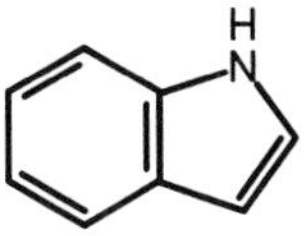 (4) 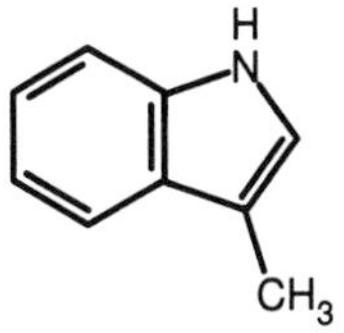 (5)

Interestingly, the 6,6′-dibromo derivative of indigo (6) occurs in marine molluscs of the genera *Murex* and *Nucella*, and a few related whelks. This red-violet pigment, known as Tyrian purple, was used for centuries by nobility as a rare, prestigious, and hence very costly dye. The compound is of interest biochemically since brominated compounds are seldom found in nature.

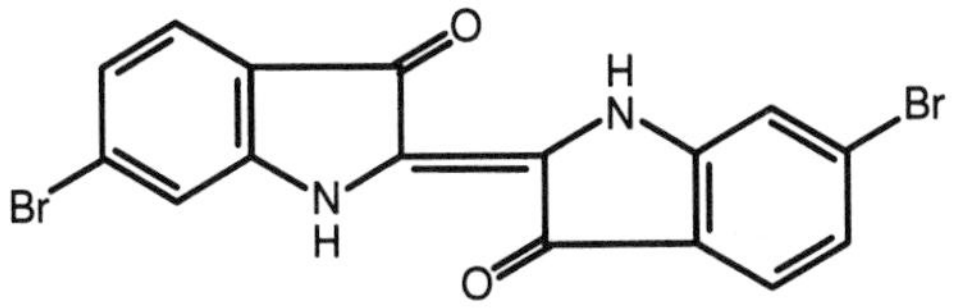

(6)

Indole 3-acetic acid (7) is recognized as the principal natural auxin (growth regulator) in plants. Tryptamine (8) the decarboxylation product of tryptophan is also found in plants, as is serotonin (5-hydroxytryptamine; 9) which occurs in bananas, plums, pineapple and nuts. Found in the tissues and fluids of vertebrates and invertebrates, serotonin plays a role in a variety of physiological functions. It is a vasoactive compound, stored in thrombocytes and mast cells of the blood. It serves as a neurotransmitter, stimulates peristalsis of the intestine and induces vasoconstriction or vasodilation of the respiratory tract and vascular system.

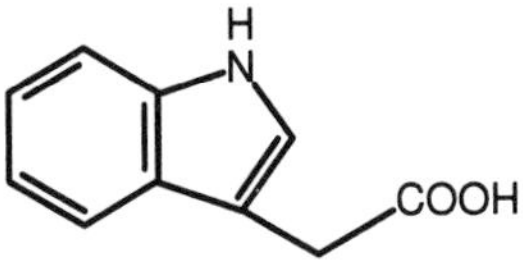

(7)

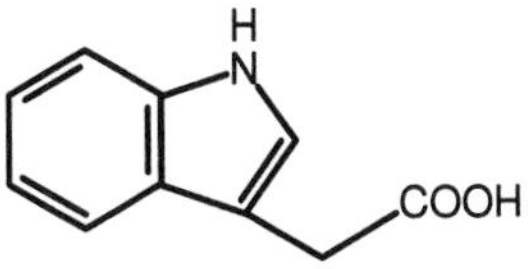

(8) 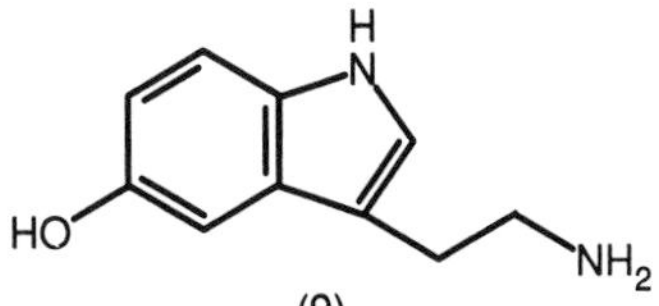 (9)

The 4-hydroxy analogue of serotonin is the active principle of an ancient Mexican drug prepared from mushrooms belonging to the genus *Psilocybe*. Pharmacologically active derivatives from the same source include psilocin (10) and psilocybin (11) which have potent hallucinogenic properties.

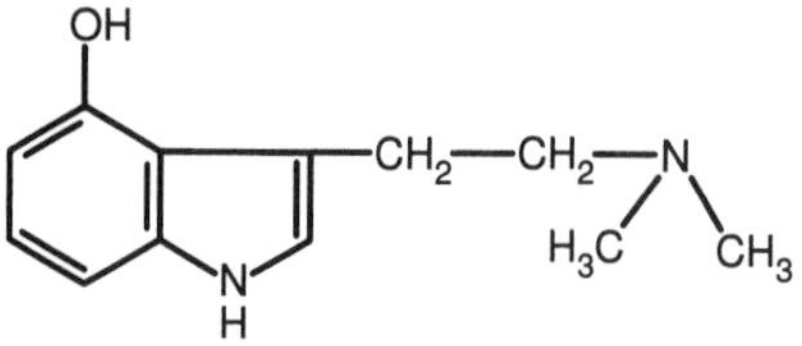

(10)

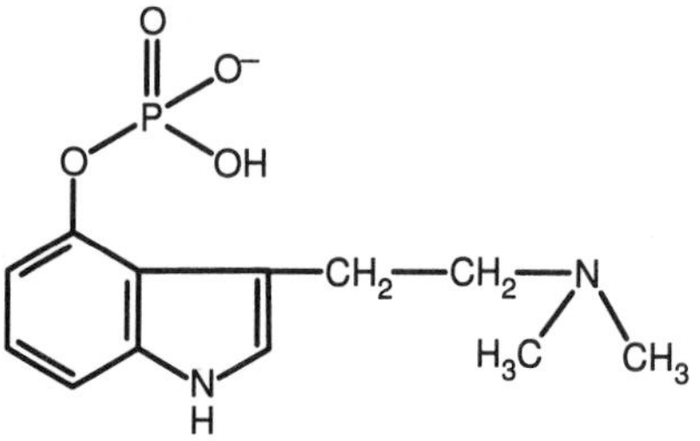

(11)

9.2 CHEMICAL PROPERTIES OF BIOCHEMICAL INTEREST

Indole (4) has an aromatic ring system of ten π-electrons. Being a benzopyrrole, and as with pyrrole itself, delocalization of the lone pair of electrons of the nitrogen atom is necessary for aromaticity. This makes the lone pair unavailable for protonation in mildly acidic conditions and, like pyrrole, indole is therefore weakly basic. Also like pyrrole, indole undergoes facile electrophilic substitution but whereas pyrrole reacts primarily at the 2- and 5- positions, indole preferentially reacts at the 5-position.

A wide variety of reagents, including atmospheric oxygen, oxidize indoles, usually attacking the 3-position. Indole, itself, is photo-oxidized by atmospheric oxygen to 3-oxoindole which, by radical coupling, dimerizes to indigo (2).

9.3 BIOSYNTHESIS AND INTERCONVERSION

The indole ring system is biosynthesized in the form of the amino acid tryptophan (1). Tryptophan is, however, an 'essential' amino acid which means that it cannot be synthesized in the human body but, like the vitamins, must be supplied in the diet. Biosynthesis of tryptophan is, then, a matter of plant or microbial biochemistry. The starting point is anthranilic acid and the first step in the biosynthesis requires the biological ribotylating agent 5-phosphoribosyl-1-pyrophosphate (PRPP), a compound involved in purine and pyrimidine nucleotide formation and salvage. It is of interest that PRPP is also required in the first step in the biosynthesis of the imidazole amino acid histidine (Chapter 2). In tryptophan biosynthesis, anthranilate is *N*-ribotylated by PRPP (Fig. 9.1) to yield *N*-5′-phosphoribosylanthranilate. This undergoes an Amadori rearrangement in which the ribose moiety is converted to ribulose. Decarboxylation is followed by ring-closure to give indole-3-glycerol phosphate. Then, in a reaction catalysed by the enzyme tryptophan synthase and requiring pyridoxal 5′-phosphate as coenzyme, the glycerol phosphate side-chain on the indole is replaced by alanine in a condensation reaction with 3-hydroxyalanine, i.e. serine (Fig. 9.1).

Tryptophan is converted into a number of other indole derivatives, especially in plants where it is the starting point for the synthesis of a vast family of indole alkaloids. More than 1200 of these are known and whereas some are relatively simple derivatives of indole, the majority are of complex structure, mostly consisting of an indole nucleus and a C_9- or C_{10}-monoterpene moiety. Examples include strychnine (12), yohimbine (13), and the lysergic acid derivative ergotamine (14). It is beyond the scope of this book to survey the multifarious biosynthetic

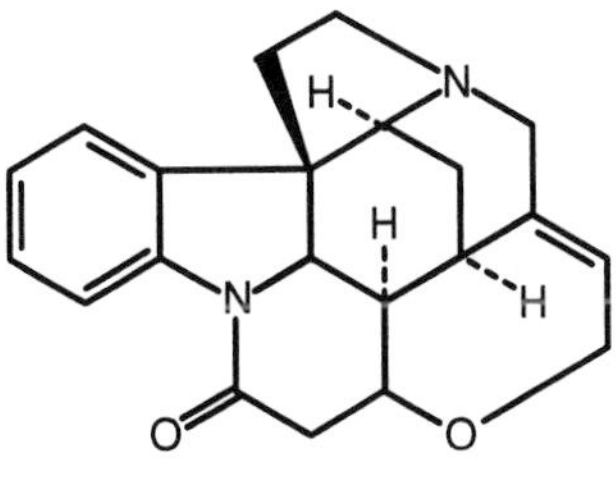

(12)

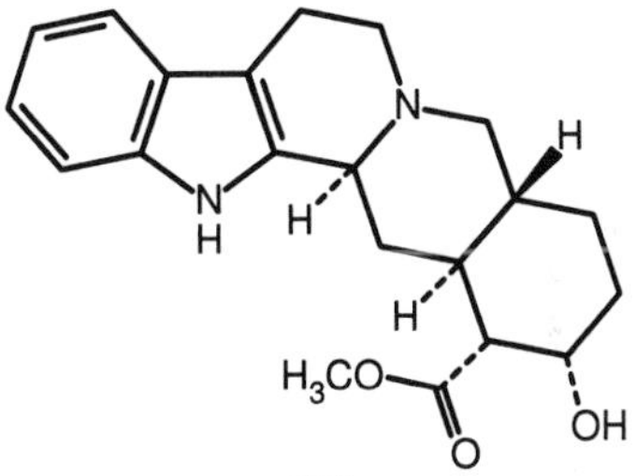

(13)

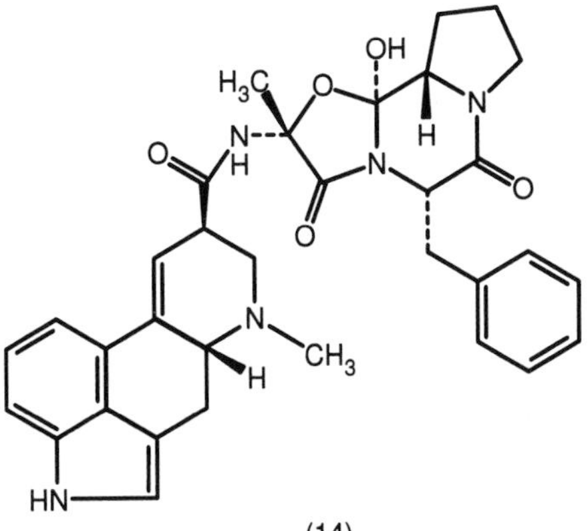

(14)

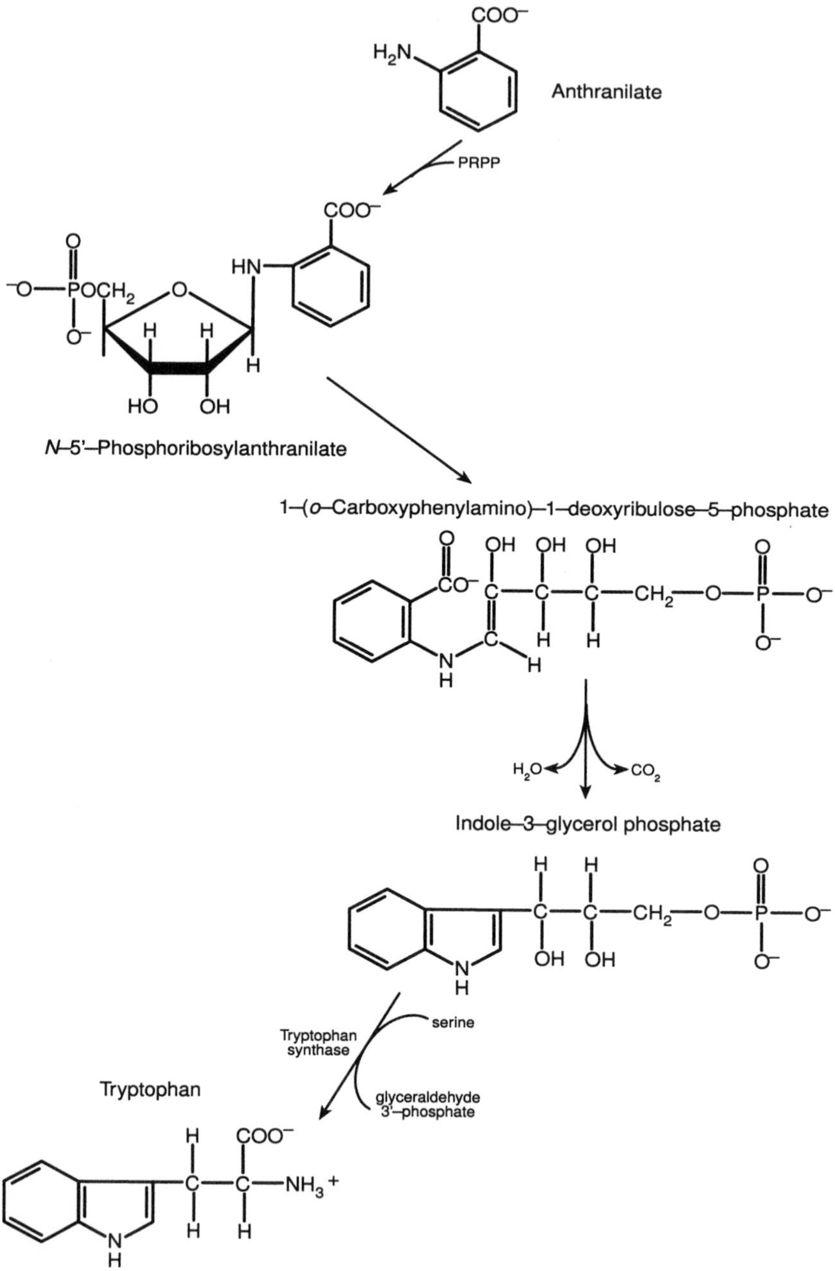

Anthranilate

PRPP

N-5'-Phosphoribosylanthranilate

1-(o-Carboxyphenylamino)-1-deoxyribulose-5-phosphate

H_2O CO_2

Indole-3-glycerol phosphate

Tryptophan synthase

serine

glyceraldehyde 3'-phosphate

Tryptophan

Fig. 9.1 Biosynthesis of tryptophan from anthranilic acid. Anthranilate is ribotylated by 5-phosphoribosyl-1-pyrophosphate (PRPP). The ribotide then undergoes an Amadori rearrangement followed by decarboxylation and ring-closure. Finally, the glycerol phosphate side-chain is replaced by alanine, supplied by a molecule of serine.

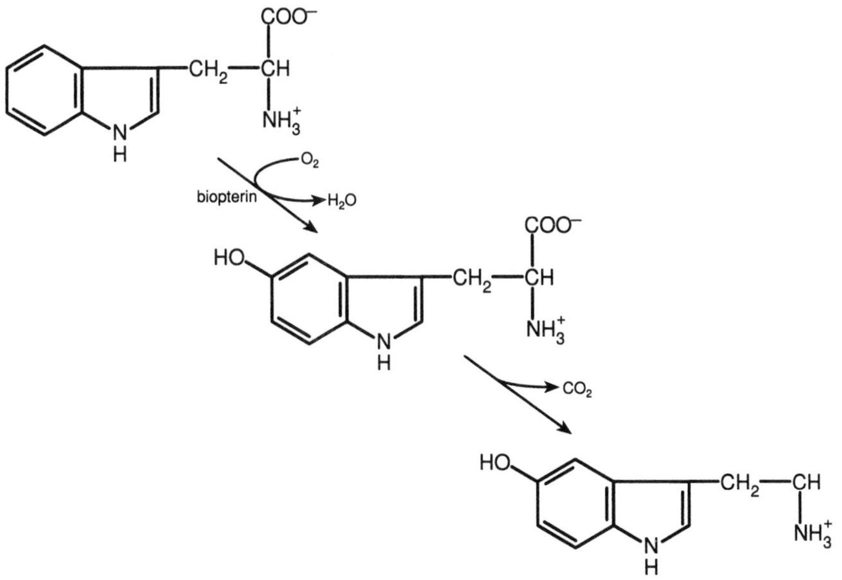

Fig. 9.2 Formation of serotonin (5-hydroxytryptamine) from tryptophan. Tryptophan is hydroxylated by a biopterin-dependent hydroxylase and then enzymically decarboxylated.

processes by which these compounds arise and the interested reader is referred, instead, to the specialist works on alkaloids listed at the end of this chapter.

Hydroxylation of tryptophan (Fig. 9.2) occurs in plants, animals, and microorganisms, and is catalysed by a biopterin-dependent enzyme similar to the phenylalanine hydroxylase described in Chapter 4. The product, 5-hydroxytryptophan, is enzymically decarboxylated in biological systems (Fig. 9.2) to yield 5-hydroxytryptamine (serotonin) which has important physiological and pharmacological activity in animals. In plants, direct decarboxylation of tryptophan also occurs, i.e. without prior 5-hydroxylation, and this gives rise to tryptamine (Fig. 9.3) which can be oxidatively deaminated to yield the plant growth hormone indole 3-acetic acid (auxin). In most plants, the main route from tryptophan to indole 3-acetic acid appears to be *via* indole pyruvic acid, formed by the action of tryptophan aminotransferase (Fig. 9.3). Indole 3-pyruvic acid is a very unstable compound and indole 3-acetic acid has been found amongst its chemical breakdown products. Production of indole 3-acetate from indole 3-pyruvate is facilitated in plants by a decarboxylase that forms indole 3-acetaldehyde. This aldehyde is readily oxidized, enzymically, to yield indole 3-acetic acid (Fig. 9.3).

Tryptamine and 5-hydroxytryptamine can also be methylated by a

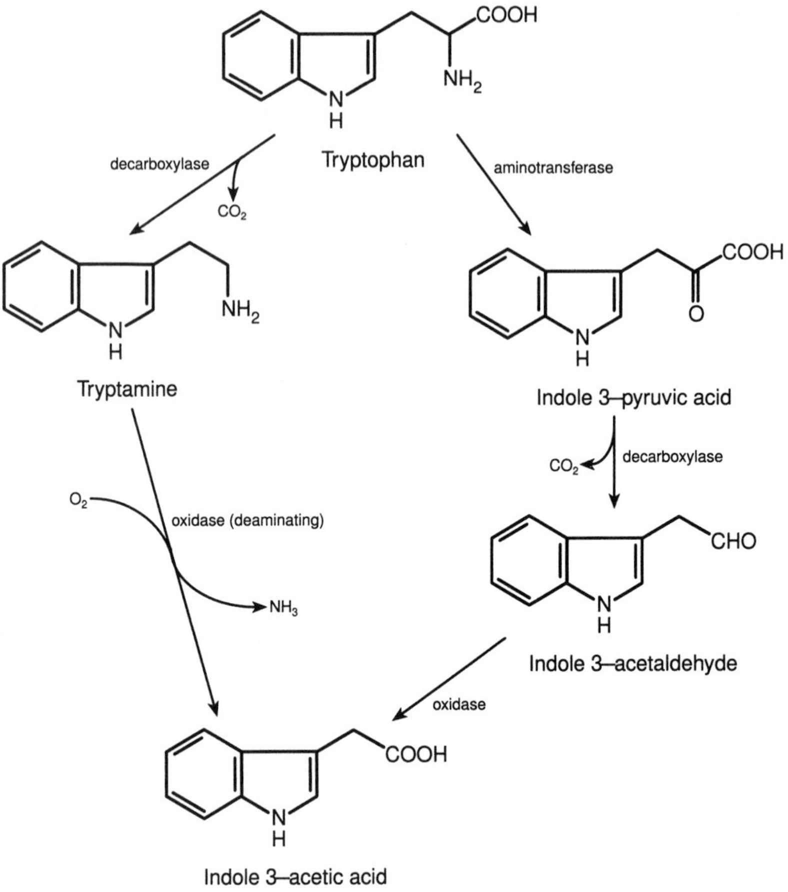

Tryptophan

decarboxylase CO_2 Tryptamine

aminotransferase Indole 3–pyruvic acid

CO_2 decarboxylase Indole 3–acetaldehyde

O_2 oxidase (deaminating) NH_3

oxidase Indole 3–acetic acid

Fig. 9.3 Formation of indole 3-acetic acid from tryptophan by plants. The main route in most plants is *via* indole 3-pyruvic acid which is formed by the action of an aminotransferase on tryptophan. Indole 3-acetic acid can also be produced by decarboxylation of tryptophan, followed by oxidation of the product, tryptamine.

number of plants, forming compounds such as gramine (15) and the hallucinogen bufotenine (16).

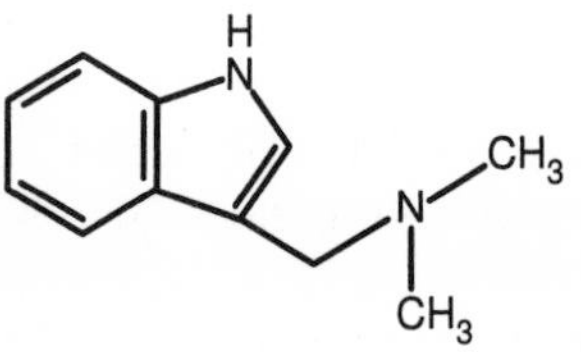

(15)

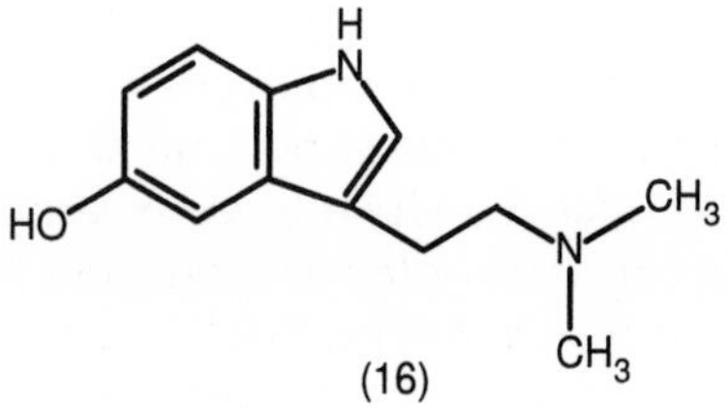

(16)

9.4 CATABOLISM

Biochemical degradation of the indole ring system is illustrated by the catabolism of tryptophan (Fig. 9.4). In the first step of this process, the pyrrole ring of the indole nucleus is cleaved by tryptophan 2,3-dioxygenase to yield *N*-formylkynurenine. Hydrolytic removal of the formyl group followed by enzymic elimination of alanine leaves 3-hydroxyanthranilate which is then oxidatively ring-opened by a monoxygenase to produce 2-amino-3-carboxymuconate semialdehyde. After enzymic decarboxylation, the semialdehyde is degraded to acetyl-CoA. An alternative fate of muconate semialdehyde, however, is ring-closure to form quinolinate which is the precursor of the pyridine ring of the coenzymes NAD and NADP (Chapter 4).

9.5 BIOCHEMICAL FUNCTIONS

Like other of the *N*-heterocyclic amino acids described in this book, tryptophan is found at the catalytic site of a number of enzymes. One such example is lysozyme, an enzyme that lyses bacterial cell walls. These structures consist, primarily, of a peptidoglycan, the carbohydrate component of which is a copolymer of *N*-acetylglucosamine and *N*-acetylmuramic acid. The sugar residues are linked through $\beta(1\rightarrow4)$ glycosidic bonds, and it is the linkage between *C*-1 of one *N*-acetylmuramic acid residue and *C*-4 of an adjacent *N*-acetylglucosamine unit that is split by lysozyme. During this process, the substrate polymer is held in place at the active site by hydrogen-bonding between the *C*-6 oxygen of one *N*-acetylglucosamine residue and the indole nitrogen atom of tryptophan 62 in the enzyme protein, and also between the hydroxyl at *C*-3 of the same sugar residue and tryptophan 63. Aspartate 101 is similarly engaged at the catalytic site in hydrogen-bonding to the substrate (Fig. 9.5).

In plants, the indole derivative indole 3-acetic acid (1) plays an important role in growth regulation. The discovery of this plant hormone can be traced back to observations made by Charles Darwin, and others, at the end of the 19th century, and later, to the studies of Went in the 1920s. It was noted that the phototropism, or tendency to grow towards a light source, exhibited by seedlings was due to migration of a diffusible compound from the light-sensitive shoot tip to a lower region of the shoot, causing rapid growth on the side of the shoot furthest from the light. The resulting unilateral growth acceleration has the effect of making the shoot arch towards the light source and is caused by cell-elongation in the growth-stimulated region.

The effects of indole 3-acetic acid in the plant vary considerably with the nature of the target tissue. It plays a major role in suppressing for-

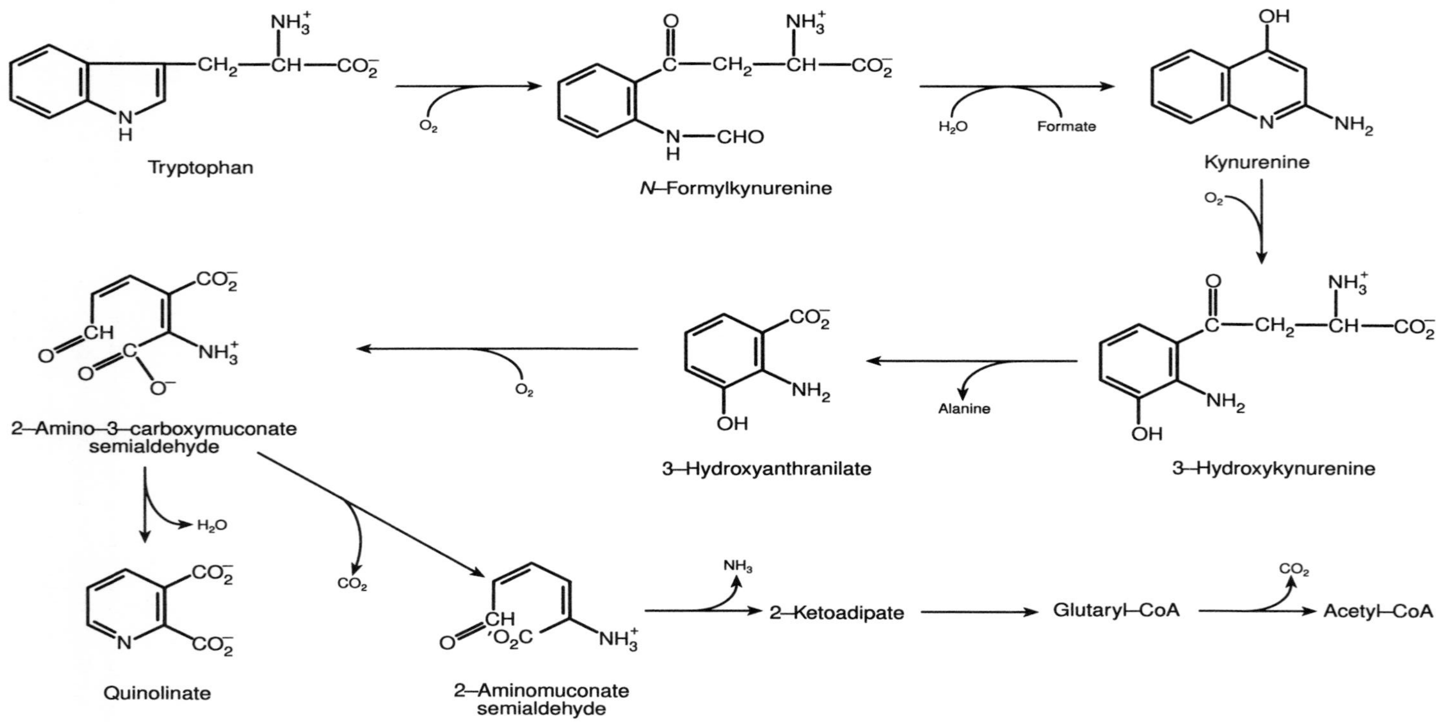

Fig. 9.4 Catabolism of tryptophan. At the 2-amino-3-carboxymuconate semialdehyde stage of the degradation, the pathway divides. The main catabolic route results in production of acetyl-CoA but the semialdehyde intermediate can also spontaneously convert to quinolinic acid, the precursor of NAD and NADP.

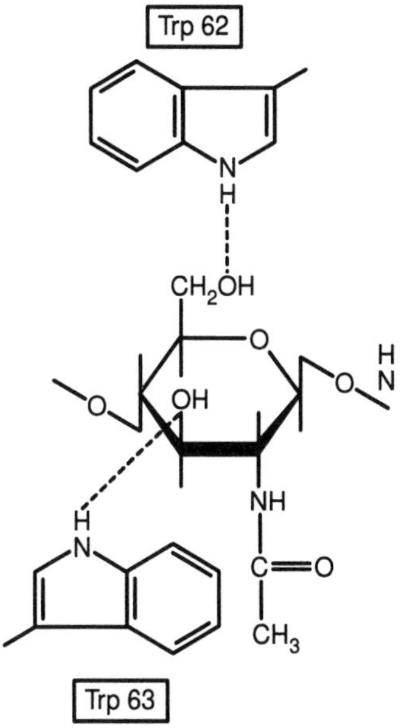

Fig. 9.5 At the catalytic site of lysozyme, an *N*-acetylglucosamine residue of the peptidoglycan component of a bacterial cell wall is held for processing. This involves hydrogen-bonding between the C-6 oxygen atom of *N*-acetylglucosamine and the indole-nitrogen atom of tryptophan residue 62 of the enzyme protein. Also involved is hydrogen-bonding between the C-3 oxygen atom of the *N*-acetylglucosamine residue and the indole-nitrogen atom of tryptophan residue 63.

mation of side shoots and in the development of fruits. In the cambium and in plant tissue cultures, it positively affects mitotic activity. It is involved in the differentiation of xylem and phloem cells, and in excised shoots it initiates development of adventitious roots. There have been reports in recent years of the isolation of receptor proteins for indole 3-acetic acid in plants but as yet no specific biochemical activity has been linked to them.

9.6 BIOLOGICAL AND PHARMACOLOGICAL ACTIVITY

More than one thousand indole alkaloids occur naturally and encompass a wide diversity of both structural complexity and pharmacological activity. Most are plant products but some are produced as a result of microbial action. A number of indoles have been used medicinally for centuries, often in the form of crude herbal extracts, and they constitute the active principles of many folk remedies from various geographical regions. Recognition of their medicinal value has led during the last 50 years to the synthesis, industrial production, and clinical use of a substantial range of novel analogues.

Over 180 years ago, the indole alkaloid strychnine was one of the first alkaloids of any type to be isolated in a pure state. It occurs in relative abundance in the seeds of *Strychnos nux-vomica* and related species and is a powerful poison. Now chiefly used in poison bait for rodents, it has been prescribed in veterinary medicine as a tonic and stimulant. This apparent pharmacological paradox is not uncommon in relation to the biological activity of alkaloids. Morphine, for example, given to man or dog, induces sleep but to horse or cat may bring on uncontrollable excitement and subsequent death.

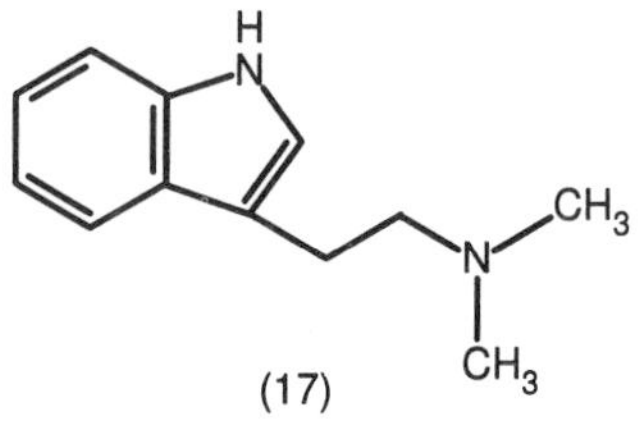

(17)

N,N-dimethyltryptamine (17) is another naturally occurring indole with pronounced physiological activity. It is a strong hallucinogen, found in several plant preparations of known psychedelic activity. These include the South American snuff *cohoba* and its West Indian equivalent *yopo*. The psychedelic effects produced by *N,N*-dimethyltryptamine involve alterations in visual perception or true hallucination and are often accompanied by euphoria and behavioural excitability. The 4-hydroxy derivative of *N,N*-dimethyltryptamine has the trivial name *psilocin* (10) and is also a psychedelic drug. *Psilocybin* (11), the naturally occurring phosphate ester of psilocin, intrinsically has much lesser activity in this respect but it is readily hydrolysed *in situ* to release psilocin. In addition to its 4-hydroxy derivative, *N,N*-dimethyltryptamine also has a pharmacologically active 5-hydroxy derivative, known as bufotenine (16). This is found in the poisonous secretions from the skin and parotid glands of toads (Bufonidae) and in some species of fungi. It increases the blood pressure and heart-rate, has a paralysing effect on the motor centres of the brain and the spinal column, and like psilocin, it is one of the psychedelic active principles of so-called 'magic mushrooms'.

Another group of indole alkaloids with a long pharmacological history are the ergot alkaloids produced by various species of the fungal genus *Claviceps* which parasitize rye, and wild grasses. In Europe during the middle ages, great epidemics occurred as people were poisoned by eating rye bread made from infected grain. An epidemic in France in 994 is described as having killed about 40,000, and even today small outbreaks are reported from time to time. The poison syndrome, known as *ergotism*, is characterized by tingling of the fingers, vomiting, diarrhoea and within a few days in severe cases, gangrene appears in the toes and fingers. Often, epileptic-like convulsions, known for centuries as 'St Anthony's fire', develop; the gangrene leads to loss of limbs.

Having the ergoline ring system in common (18) more than 30 ergot alkaloids are known. They are exemplified by ergotamine (19), ergometrine (ergovinine) (20) and lysergic acid (21). The obstetric use of ergot, which consists of the dried sclerotia (spore-containing bodies) of *Claviceps purpurea*, was known in the sixteenth century and appeared in the *London Pharmacopoeia* in 1836. Today, pure preparations of individual ergot alkaloids, most commonly ergotamine or a synthetic analogue, are still used as an oxytocics to induce contraction of the smooth muscle of the wall of the uterus. The effect of ergotamine in vasoconstricting peripheral arterioles has been made use of in treating migraine. Ergometrine is also used as an oxytocic. Lysergic acid and

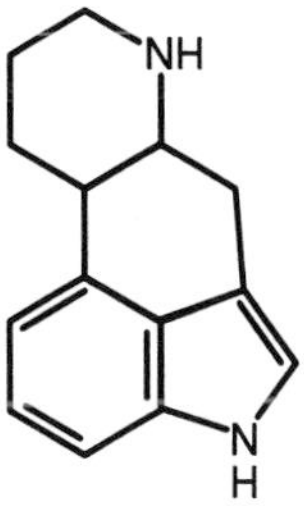

(18)

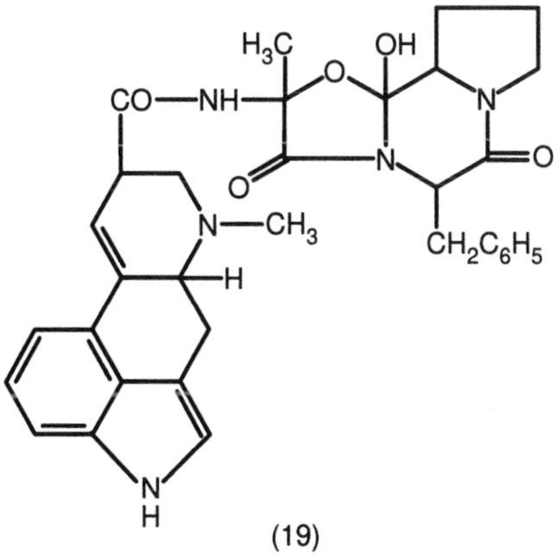

(19)

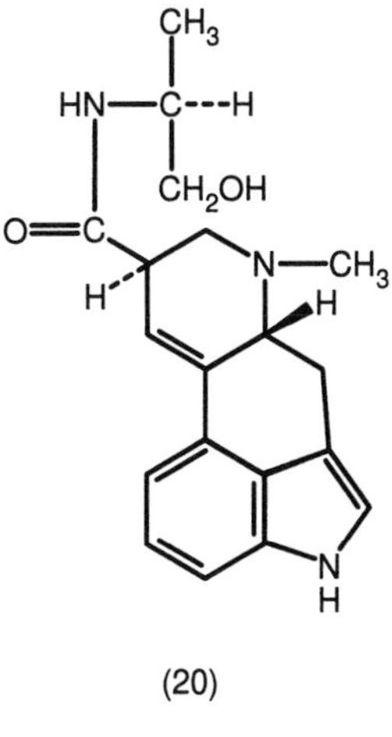

(20)

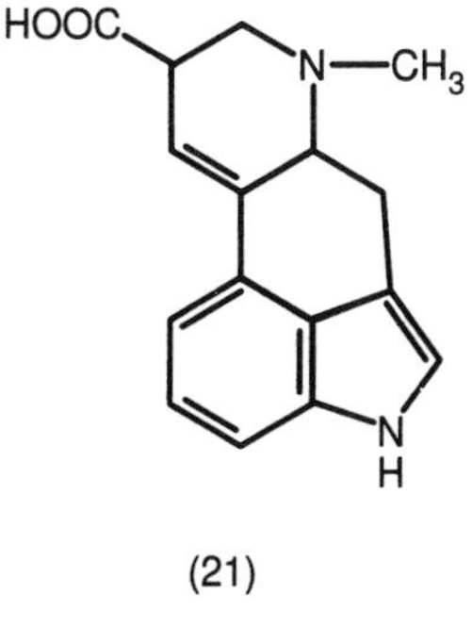

(21)

especially its synthetic diethylamide (LSD) are well-known as hallucinogens.

A group of around 20 dimeric indole alkaloids have been isolated from the periwinkle *Catharanthus rosea*, formerly known as *Vinca rosea*. These 'vinca' alkaloids possess antineoplastic activity and are now extensively used in cancer chemotherapy. Vincristine (22) and vinblastine (22) are the best known and most widely used of these indole alkaloids. Vinblastine is especially valuable in the treatment of Hodgkins disease (lymphadenoma) and chorionepithelioma, whereas vincristine is used principally in the treatment of leukaemia in children. As with most of the drugs currently used to treat cancers, the cytotoxicity upon which their clinical value rests is selective in that they attack rapidly dividing cells, but not specific in that normal cells are also affected to a lesser extent. This effect on normal cells is the cause of the unpleasant side-effects of cancer chemotherapy.

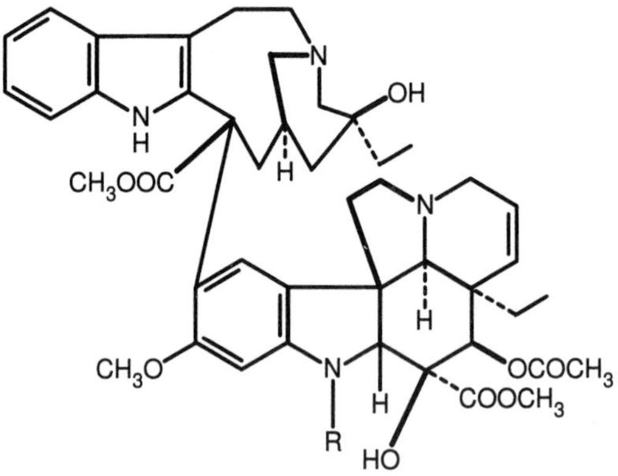

Vincristine, R = CHO Vinblastine, R = CH_3

(22)

9.7 SYNTHETIC COMPOUNDS OF CLINICAL IMPORTANCE

Turning now to synthetic indoles of medicinal interest, one of the best known is probably indomethacin (23) an anti-inflammatory drug com-

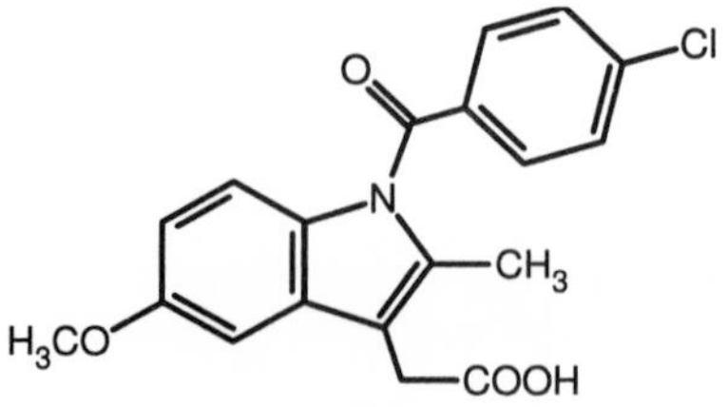

(23)

monly used in treating osteoarthritis and gout, and also of some value in managing rheumatoid arthritis. It is one of the most significant and widely prescribed non-steroidal anti-inflammatory agents. Cinmetacin (24) a closely related compound is also active as an anti-inflammatory compound. Oxypertine (25) is a clinically useful tranquillizer; Iprindole (26) is an antidepressant, functioning by blocking the uptake of catecholamines and serotonin by adrenergic neurons; and Pindolol (27) is a β-adrenergic blocking drug (a β-blocker). In the latter case, however, it is doubtful that the indole ring is important since the prototypical β-blocker propranolol is isosteric with it but has a naphthalene ring in place of indole.

The notorious drug LSD (lysergic acid *N,N*-diethylamide) (28) is a synthetic derivative of lysergic acid. It is the most potent psychotomimetic compound known and induces a state resembling schizophrenia.

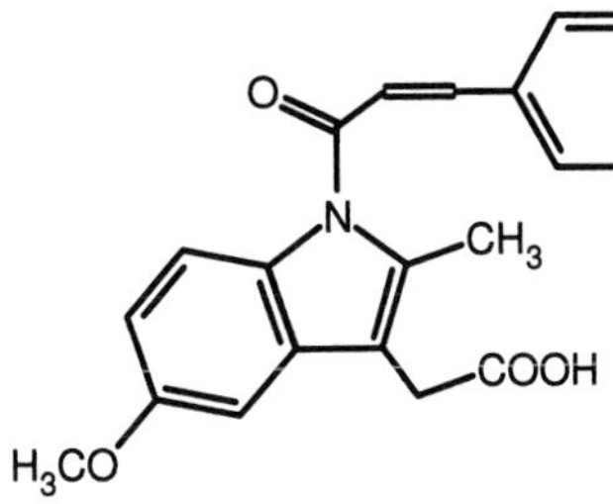

O N CH_3 COOH H_3CO

(24)

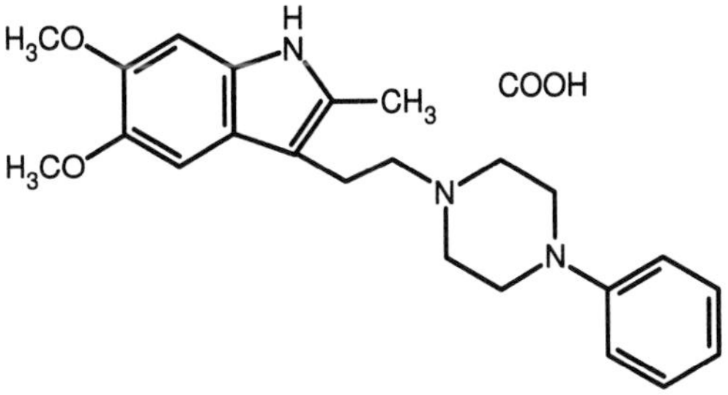

H N H_3CO H_3CO CH_3 COOH N N

(25)

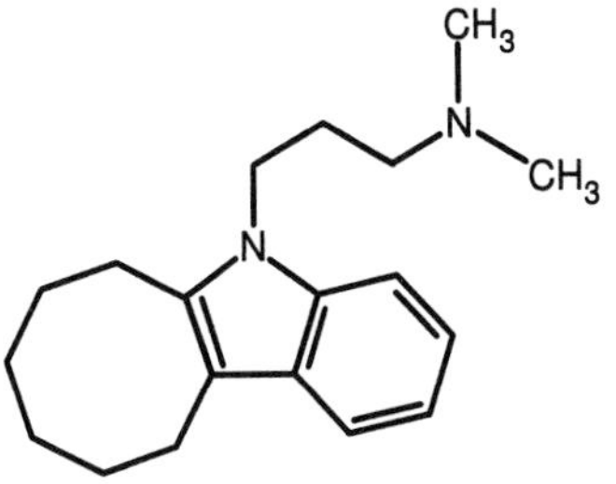

CH_3 N CH_3 N

(26)

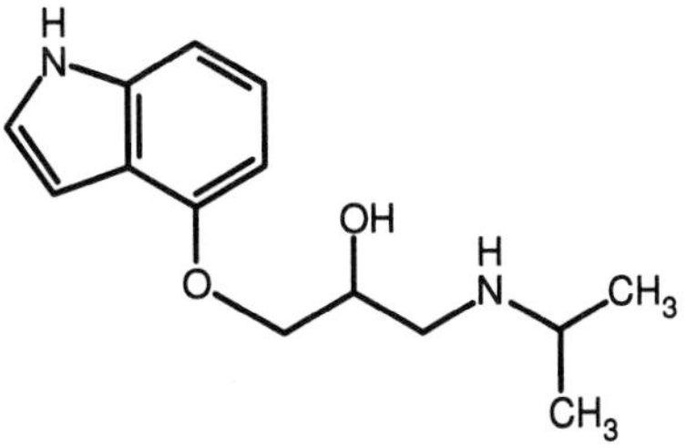

H N OH O H N CH_3 CH_3

(27)

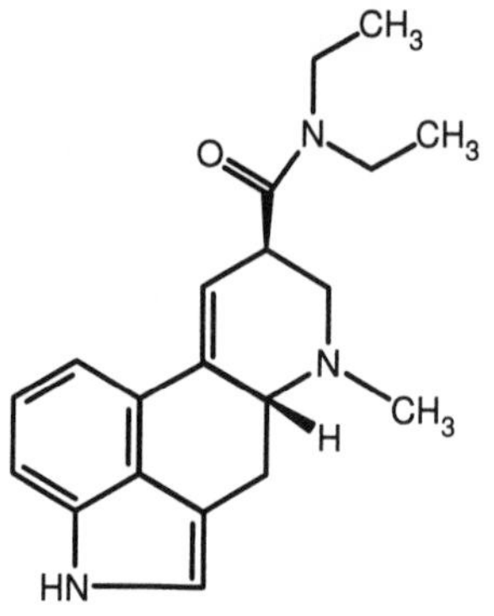

(28)

Discovered during research into the pharmacological effects of ergot alkaloids, its powerful hallucinogenic properties were not realized for five years, and followed accidental ingestion of a contaminating trace of the compound. Subsequently it has been shown that LSD interferes with the neurotransmitter action of serotonin.

9.8 COMPOUNDS OF AGRICULTURAL AND HORTICULTURAL IMPORTANCE

The pronounced physiological effects of externally applied indole 3-acetic acid (auxin) on plant growth and development has been exploited in both agriculture and horticulture. A number of synthetic auxins have been developed but two of these have found most favour. They are 2,4-dichlorophenoxyacetic acid (2,4-D) and naphthalene 1-acetic acid (NAA); their structures are shown in Fig. 9.6. Because its effects are pri-

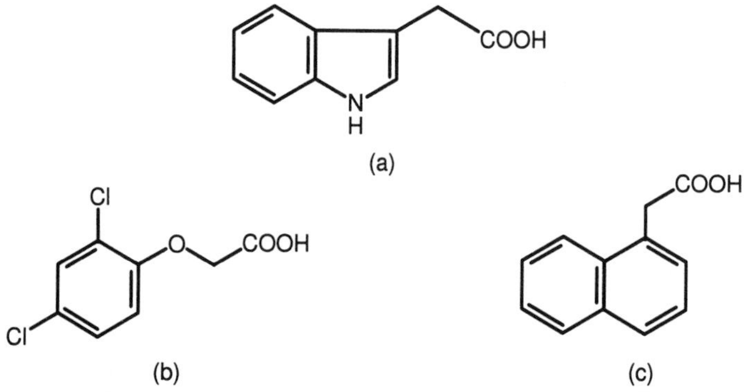

(a)

(b) (c)

Fig. 9.6 (a) The plant growth hormone indole 3-acetic acid and its two synthetic analogues, (b) 2,4-dichlorophenoxyacetic acid (2,4-D) and (c) naphthalene 1-acetic acid (NAA). The analogues (b) and (c) are commonly used in agriculture and horticulture.

marily on broad-leaf plants, 2,4-D is used as a selective herbicide. It works, in effect, by stimulating weed seedlings to out grow themselves. Another commercial use of 2,4-D is in rubber plantations where its application induces old rubber trees to produce more latex. NAA is mainly used in 'rooting powders' to stimulate the development of adventitious roots by stem and leaf cuttings, a process of great importance in the large scale propagation of plants commercially.

WIDER READING (BOOKS AND REVIEWS)

Brown, R.T. and Joule, J.A. (1979), in *Heterocyclic Chemistry* (ed P.G. Sammes), Vol. 4 of *Comprehensive Organic Chemistry* (ed D. Barton and W.D. Ellis), Pergamon Press, Oxford.

Cordell, G.A. (1981) *Introduction to Alkaloids*, Wiley-Interscience, New York.

Dalton, D.R. (1979) *The Alkaloids: The Fundamental Chemistry*. Marcel Decker, New York.

Gilchrist, T.L. (1997) *Heterocyclic Chemistry*, 3rd edn, Longman, Harlow, U.K.

Houlihan, W.J. (ed) (1972) *Indoles*, in *The Chemistry of Heterocyclic Compounds*, (eds A. Weissburber and E.C. Taylor), Vol. 25, Parts 1–3, Wiley-Interscience, New York.

Phillipson, J.D. and Zenk, M.H. (eds) (1980) *Indole and Biogenetically Related Alkaloids*, Academic Press, London.

Saxton, J.E. (ed) (1979) *Indoles*, in *The Chemistry of Heterocyclic Compounds*, (eds A. Weissburger and E.C. Taylor), Vol. 25, Part 4, Wiley-Interscience, New York.

CHAPTER 10

N-Heterocycles and living systems – conclusions

The extraordinary frequency with which *N*-heterocyclic compounds occur as essential components of biochemical pathways inevitably poses the question as to what it is about these ring structures that singles them out by nature as biochemical tools. In considering this question, we could usefully ask what the various roles planed by these compounds have in common. The answer has to be 'catalysis'.

In most cases, it is self-evident when a heterocycle is operating as a biochemical catalyst, e.g. the nucleotides NAD, NADP and FAD all clearly function as redox catalysts. It is, however, a little less obvious that the nucleotide heteropolymers, the nucleic acids, also play catalytic roles. They function primarily as molecular templates for the assembly of other polymers, e.g. other nucleic acids or proteins. In eukaryotes these templates have half-lives ranging from 1 to 24 hours, and as they are re-used many times before becoming redundant, they can be regarded as catalysts in nucleic acid or protein biosynthesis. In 1983, an even more obvious example of nucleic acids behaving catalytically came to light when it was shown by Altman and his co-workers that the enzyme ribonuclease P, which cleaves specific RNA molecules to yield functional tRNAs, is a ribonucleoprotein (i.e. a conjugate of RNA and protein) and that its catalytic activity resides in the RNA moiety and not in the protein. Other examples of RNA behaving enzymically in this way have since been discovered and have led to the generic designation 'ribozyme' to describe them.

What is there about *N*-heterocyclic ring systems that has singled them out, as a group, for biocatalysis? There are a number of answers to this question. To begin with, *N*-heterocycles are capable of behaving as acids or bases. Those like pyrrole with a ring imino group (–NH–) behave as acids, albeit often weak acids. Others with pyridine-like nitrogen atoms (=N–) behave as bases. Compounds, like imidazole, with two ring nitrogen atoms, are amphoteric. These properties are of prime importance in general acid-base catalysis. Pyridine, on the other hand, being a π-deficient ring system, is easily reduced and makes a good redox catalyst, as seen with the pyridine nucleotides NAD and NADP.

Similarly, the isoalloxazine ring system of riboflavin is easily and reversibly reduced, and functions biochemically in the form of FMN and FAD in other redox reactions. The isoalloxazine nucleus of riboflavin has another useful property in relation to its biological function. Whereas it is a two-electron acceptor, it can be reduced in two single-electron steps with the intermediate formation of a semiquinone free radical. Conversely, reduced flavins can be readily reoxidized by one-electron acceptors.

N-Heterocyclic compounds have a ready capability of forming coordination complexes with metal ions, usually cations. There is a range of biochemical catalysts of this type. Haem, for example, is a stable complex of photoporphyrin IX and Fe(II). With the cytochromes, the haem complex undergoes reversible oxidation-reduction at the iron atom and thereby functions as an electron transporter. With the respiratory pigment haemoglobin, the haem complex acts as an oxygen-carrier without itself being oxidized in the process. Chlorophyll consists of a pheoporphyrin coordination complex with Mg^{2+}. In the molecular structure of coenzyme B_{12}, a pseudoporphyrin (*corrin*) is in a coordination complex with a Co ion; this can be in the I, II or III oxidation state depending on the reaction in which it is involved. This heterocyclic coenzyme is also noteworthy in that, at the present time, it affords the only known example of a covalent carbon-metal bond in a naturally occurring compound.

Another facet of the chemical, and hence biochemical, versatility of many *N*-heterocyclic compounds is their tautomerism. This exists where a compound has in addition to a ring nitrogen atom, a substituent that offers an alternative site for protonation. Biologically, these substituents are most commonly =O or =S. In aqueous solution, rapid intramolecular proton transfer between the nitrogen and the substituent results in establishment of a pH-dependent equilibrium between the tautomers. Except at a pH approximating to a pK_a, one tautomeric form predominates. Such equilibria are particularly important with purine and pyrimidine derivatives. An example, that of the pyrimidine uracil is shown in Fig. 10.1. At the neutral pH conditions found in most biological systems, it is the oxo tautomer of uracil that predominates. A similar situation is seen with the other purines and pyrimidines. It is the oxo

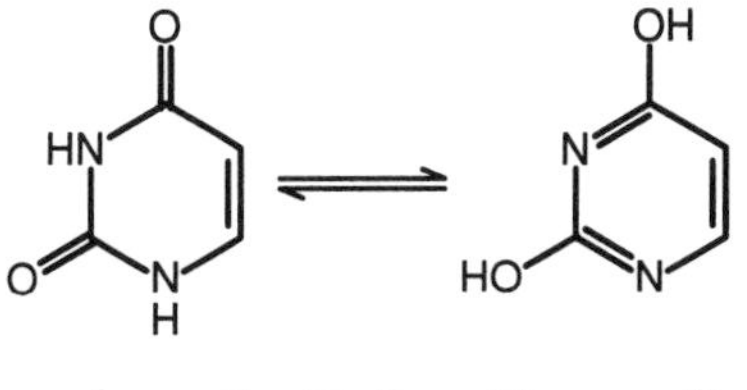

Fig. 10.1 Tautomerism of uracil. At the pH prevailing in most biological systems, the oxo tautomer (left) predominates.

tautomeric form that facilitates the specific hydrogen-bonding seen between base pairs and which stabilizes the double helical 3-D structure of DNA (Fig. 10.2). Because of the relative molecular dimensions and

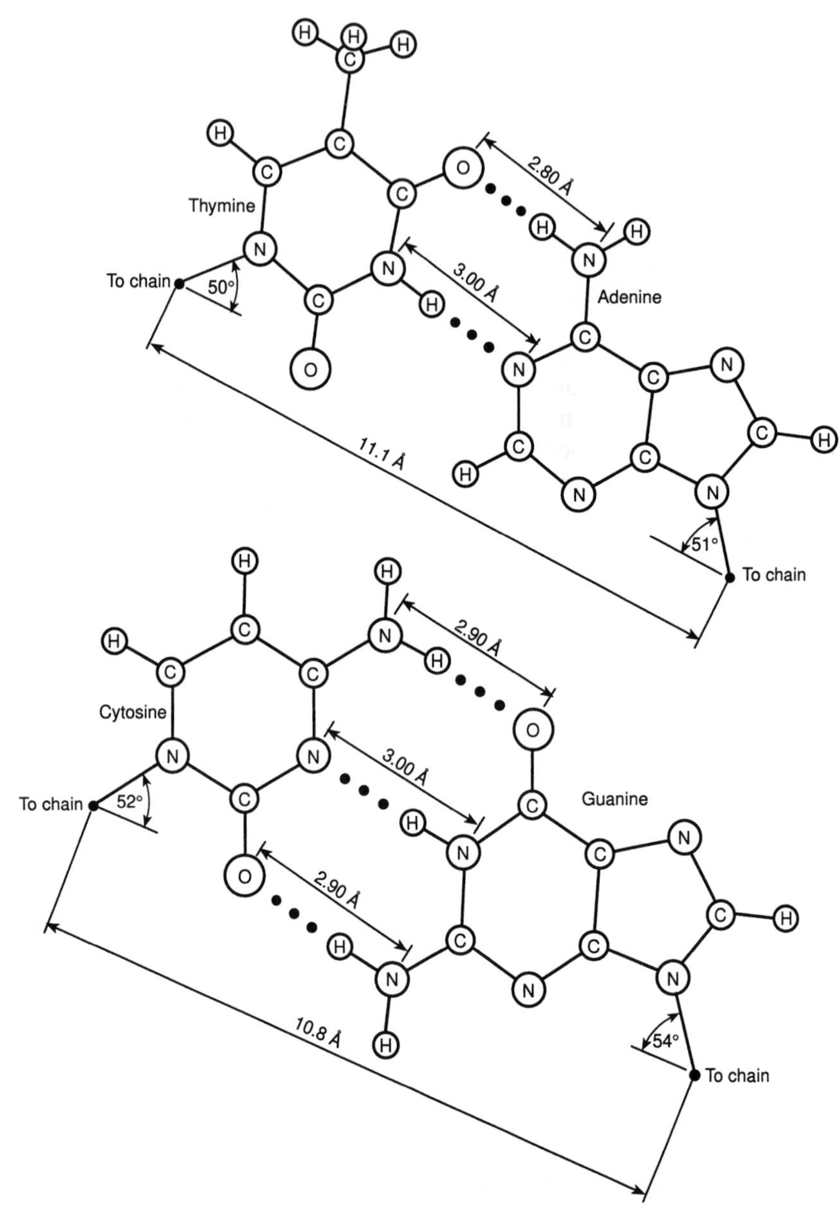

Fig. 10.2 The specific hydrogen-bonding between the base pair thymine and adenine, and between cytosine and guanine, which stabilizes the double helical structure of DNA. Each of the pairs shown are, in effect, seen in a transverse section through the double helix.

orientation, such bonding in DNA occurs specifically between adenine and thymine residues opposing each other in the complementary strands, and between guanine and cytosine residues similarly opposing each other in the double helical structure.

Intramolecular hydrogen-bonding is also important in enzymic catalysis and often occurs between the *N*-heterocyclic R group of a contact amino acid residue in the protein and a functional group on the substrate molecule. This occurs, for example, with the enzyme lactate dehydrogenase where the histidine 195 residue bonds with the substrate hydroxyl group. Similar hydrogen-bonding is also seen with the enzyme chymotrypsin, the reaction mechanism of which is shown, in detail, in Fig. 2.6.

Electrostatic interactions often play a significant role in catalysis and in this respect, too, *N*-heterocycles are well-placed. In addition to their ability to form ions, most *N*-heterocycles are dipoles and this can have a substantial influence on the secondary and tertiary structure of a protein. Where, for example, the terminal COO^- of a polypeptide chain is in the vicinity of a positively charged imidazolium histidine residue, the attraction could modify the conformation significantly.

That some enzymes react chemically with their substrates to form covalently bonded enzyme-substrate complexes was an important discovery in enzymology because it helped to demonstrate that the enzyme-catalysed reaction is not mechanistically different to other types of chemical reaction. Covalent catalysis involving *N*-heterocyclic compounds is not uncommon in biochemistry, and a number of *N*-heterocyclic coenzymes, together with *N*-heterocyclic amino acid residues in enzyme proteins, form covalent bonds with their substrates during catalysis. The coenzyme pyridoxal 5′-phosphate provides a good example of covalent catalysis. The first step in biochemical reactions catalysed by pyridoxal 5′-phosphate is combination of the α-amino group of the amino acid substrate with the carbonyl group of the coenzyme to form a Schiff's base (Fig. 10.3). In the absence of substrate, the carbonyl group is similarly linked with the ε-amino group of a specific lysine residue at the active site but this is displaced on arrival of the amino acid substrate (Fig. 10.3). The Schiff's base formed between the substrate amino acid and pyridoxal 5′-phosphate then undergoes the catalysed reaction.

Hydrophobic interaction, too, plays a part in some catalytic processes in which *N*-heterocyclic compounds are involved. For example, although the purine and pyrimidine base constituents of nucleic acids are essentially water-soluble, they are of relatively low solubility and the interaction with one another is stronger than their individual interactions with water. Compounds such as alcohols, ureas and carbamates, which increase the water solubility of bases tend to cause denaturation of DNA; the more the solubility is increased, the more effective the

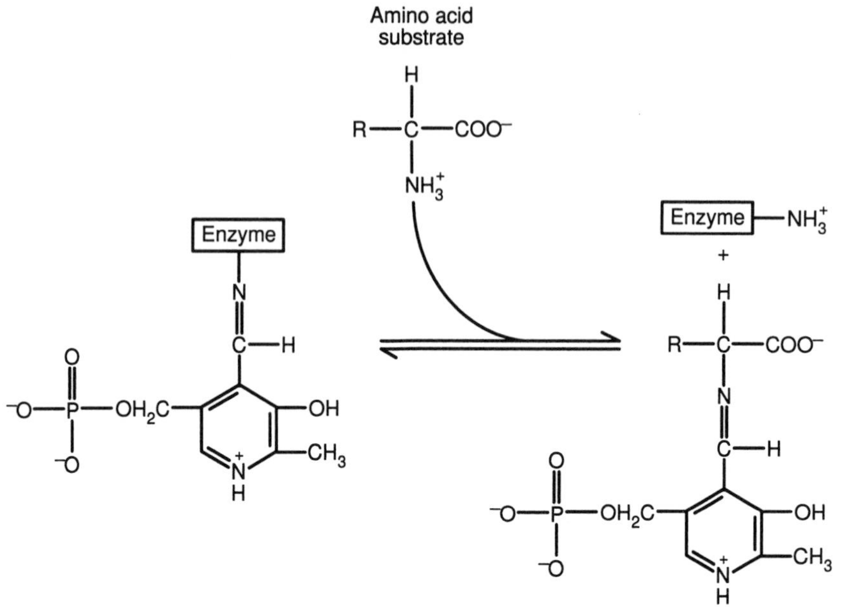

Fig. 10.3 The first step in a pyridoxal 5′-phosphate-dependent enzymic reaction. The coenzyme is linked through its carboxyl group to the amino group of a lysine residue in the enzyme protein. When the amino acid substrate appears, the lysine amino group is displaced by the substrate's amino group. The Schiff's base (right) then undergoes the catalysed reaction.

additive is in denaturing the nucleic acid. There is evidence, too, that in some enzymic reactions hydrophobic interaction increases the rate of reaction by facilitating approximation of the reactants, i.e. causing the substrate(s) and the enzymic catalytic site to come into close proximity with one another.

Consideration of the central role of *N*-heterocyclic compounds in biochemistry poses several other important but so far unanswered questions. What is it, for example, that has led to the predominance of adenine nucleotides in metabolism, and especially in its regulation and integration? The ubiquity of ATP in biochemical energy transduction processes and the frequency with which adenine nucleotides turn up as allosteric regulators of enzymic activity is patently self-evident. Several coenzymes, e.g. NAD, NADP, FAD and coenzyme A, contain within their respective structures an AMP unit. Adenosine 3′,5′-cyclic monophosphate (cyclic AMP) is a cell-signalling compound and key intermediary between many hormones and their respective biochemical effects. Adenosine is a structural component of the sulphating coenzyme 3′-phospho-adenosine-5′-phosphosulphate (PAPS), of the methylating coenzyme *S*-adenosylmethionine, and of coenzyme B_{12}.

Turning to pyrimidines, what is there about pyrimidine nucleoside diphosphates that makes them so suitable as transporting enzymes. Carbohydrate interconversion, glycosylation, and polysaccharide synthesis are heavily dependent on compounds like UDP-glucose and its sugar analogues. Phospholipid metabolism uses CDP-alcohols, like CDP-ethanolamine and CDP-choline, in a similar way, and CDP-ribitol is involved in bacterial cell wall synthesis. What is there about the biochemical and chemical properties of UDP that makes it suitable to transport aldoses whereas the alcohols favour CDP?

As we have seen, nucleotides are precursors of nucleic acids and coenzymes, they are often coenzymes in their own right, and they function both as allosteric regulators of enzymes and as cell-signalling compounds. Could there, then, be connexions between the central information system encapsulated in the nucleic acids and the other biochemical roles of purine and pyrimidine derivatives? If the answer is yes, what are these connexions?

Because purine and pyrimidine nucleotides are present in tissues in low concentrations yet are in demand for the assembly of nucleic acids and coenzymes, their individual concentrations are rate-limiting for many biochemical processes and control others through allosterism. Consequently, a sudden increase or decrease in the cellular concentration of a given nucleotide could have profound metabolic effects. This is borne out by the effect of administering D-galactosamine to rats. Intraperitoneal injection of this seemingly innocuous aminosugar induces hepatitis-like symptoms, including liver damage, attributable to the formation of UDP-galactosamine which cannot be further metabolized. Continuous production of this sugar nucleotide has the effect of sequestering all available uridine nucleotides and causing a severe depletion of this family of compounds in the liver, which is the biggest user of uridine nucleotides since it is the main seat of carbohydrate metabolism in the body. Uridine nucleotide concentrations of below 30% of the normal level cause liver damage.

It has been suggested that there could be a direct link between the nucleic acid pool and these critical concentrations of individual nucleotides in tissues. Some species of RNA contain homopolymeric base sequences, for example most eukaryotic mRNAs have polyA tails of about 250 residues and there may be other species of RNA with much longer sequences. Sudden depolymerization of them could release amounts of AMP that would be of significance within the confines of a cell compartment and so affect the metabolic balance.

In considering the selection, by nature, of *N*-heterocyclic compounds as biochemical catalysts, regulators and signalling molecules, it is appropriate to look at the availability of these compounds in the biological or *quasi* biological environment. During experimental studies concerning the origin of life on earth, Miller and his colleagues, working in the

USA in the 1950s, showed that amongst the compounds that it was possible to form by sparking or irradiating gaseous mixtures approximating to the primitive earth atmosphere, were a number of biochemically important *N*-heterocycles. These experiments were designed to simulate conditions on the primitive earth, with sparking representing lightning discharges, and irradiation with ultraviolet light imitating exposure to bright sunlight. During the studies, it was shown that adenine, uracil, pyrrole, porphyrins, and other *N*-heterocyclic compounds were formed. It is, therefore, reasonable to suppose that such compounds would have been available in the prebiotic environment and that they became incorporated into the nascent living system that evolved. The overwhelming chemical similarity of all contemporary living systems points to a common origin. Darwin's biological observations led him to a similar conclusion: he wrote in 1859 in his book *On the Origin of the Species*, 'Probably all the organic beings which have ever lived on this earth have descended from some one primordial form'.

The first biocatalysts, antecedents of today's enzymes, were very likely transition metal ions either alone or in complexes with *N*-heterocycles. The protein component would have come later as these biopolymers evolved. Support for this suggestion comes from the observation that the action of some enzymes can be simulated by metal ions alone, copper (Cu II) ions for example, catalyse the oxidation of ascorbic acid (vitamin C). Addition of a trace of egg albumin or a similar non-specific protein, greatly increases the efficiency of the copper ions in this reaction. Addition of the specific protein which, together with copper ions, constitutes the enzyme ascorbate oxidase, increases the rate of oxidation by several orders of magnitude. A similar relationship can be demonstrated between the decomposition of hydrogen peroxide by iron salts, the big stimulation of the process obtained by adding non-specific protein, and the phenomenal catalytic activity of the iron-containing protein catalase which has a k_{cat}/K_m ratio of 4×10^7 mol^{-1} s^{-1}, one of the highest enzymic activities known.

Pyridoxal, in a complex with a di- or trivalent metal ion, catalyses a variety of chemical reactions involving amino acids. The enzymes catalysing the analogous biochemical reactions do not have a metal ion requirement but hold pyridoxal, as its 5′-phosphate, in close association. It could be argued that, here, we are seeing a *N*-heterocyclic catalyst, analogous to the catalytic transition metals in the other examples, evolving into a more effective state by association with protein.

In contemporary biological systems, which we should not forget are still evolving, a vast array of *N*-heterocyclic compounds are readily available from the biosynthetic activities outlined in preceding chapters of this book. Many of these are however essentially plant or microbial products and are not synthesized by animal tissues. Where, however, such compounds are essential to animal well-being, they are obtained in the diet as vitamins or essential amino acids.

For the future, it is plain that the biochemist, the biologist and the chemist have much still to learn from the central role of *N*-heterocyclic compounds in the functioning of living systems. In the meantime, we should not forget that pursuit of this worthwhile objective, and indeed of biochemistry as a whole, rests upon the synergistic interaction of chemistry and biology. The trend away from chemistry in contemporary university courses in biology, and even in biochemistry, needs to be redressed. So, too, does the lack of biochemical input into chemistry courses.

'Where is the knowledge we have lost in information?'

T.S. Eliot

Index

Numbers in **bold** refer to page numbers of figures; those in *italics* refer to page numbers of tables.